Hands-On Science Experiments

by Deborah Crotts

Illustrated by Joseph Dolce

Notice! Copies of student pages may be reproduced by the classroom teacher for classroom use only, not for commercial resale. No part of this publication may be reproduced for storage in a retrieval system, or transmitted in any form or by any means-electronic, mechanical, recording, etc.-without the prior written permission of the publisher. Reproduction of these materials for an entire school or school system is strictly prohibited.

FS-10112 Hands on Science Experiments
All rights reserved-Printed in the U.S.A.
Copyright © 1993 Frank Schaffer Publications, Inc.
23740 Hawthorne Blvd.
Torrance, CA 90505

TABLE OF CONTENTS

CHAPTER ONE — PAGE 2
The Language and Process of Science
Classification The Scientific Method Scientific Measurement

CHAPTER TWO — PAGE 21
The Living World–Biology
Biological Organization Classifying the Living World
Microorganisms Disease Caused by Microorganisms
Insects Marine Life Mammals The Human Body
Prehistoric Plants and Animals

CHAPTER THREE — PAGE 55
Geology
Composition of the Earth Rocks and Minerals
Volcanoes and Earthquakes Mountains Groundwater Weather

CHAPTER FOUR — PAGE 76
Ecology and the Environment
Water Balance of Nature Adaptation
Conservation Pollution Recycling

CHAPTER FIVE — PAGE 97
Astronomy and Space
The Moon and the Solar System
The Sun, the Stars and the Milky Way Space

CHAPTER SIX — PAGE 116
Chemistry
The Basics: Molecules and States of Matter Chemicals
Chemical Changes Crystals Nature and Use of Chemicals
Drugs and the Human Body

CHAPTER SEVEN — PAGE 141
Physics
Force Fluid and Density Magnetism Electricity
Optics and Light Sound

Introduction

Hands-On Science Experiments is designed to augment the classroom text, to provide hands-on science activities that supplement the basic curriculum. This is in no way a comprehensive science textbook and is not designed to take the place of a text for any group of students. The activities cover a range of subjects and were chosen for student interest, minimum teacher preparation, simplicity and attainability of materials and classroom safety.

Each subject is addressed separately with Teacher Notes and activity keys preceding the student activity sheets at the front of each chapter. The Teacher Notes explain and amplify the activities. In addition, they contain suggested introductory and follow-up activities and references to other teacher resources.

The student activity sheets are self-contained, reproducible pages. The experiments require basic, readily accessible materials, and the list of materials needed for each activity is listed at the top of each student activity page. Additional information concerning the materials or preparation for the activity is also included in the Teacher Notes. Patterns are included where applicable. The majority of the experiments require minimal teacher supervision. Some of the experiments, primarily in Chapter Six: Chemistry, could pose some hazard if performed improperly. These activities should be very closely supervised as stated in the Teacher Notes.

Several of the student activity sheets are not actual science experiments; they are informational sheets to be used in conjunction with or as catalysts for further research. There are also pages designed to stimulate scientific thinking and awareness of the applications of scientific knowledge.

The general topics for each subject are listed in the table of contents under the chapter heading. Several topics could be listed under more than one discipline. Drugs, for instance, could be treated under Biology or under Chemistry, as they are in this book. Likewise, classification, which is here treated in the first chapter as part of the basic discipline of science, is often considered as a part of biology because classification is such an integral part of that branch of science. To aid in locating activities which are cross-disciplinary, a list of experiments for each chapter is provided at the front of each chapter.

CHAPTER ONE

THE LANGUAGE AND PROCESS OF SCIENCE

Teacher Notes 2

Classification 2

    Classify It 10

    Classification Teaser 11

    Classify the Critters 12

The Scientific Method 5

    The Scientific Method 13

    Formal Experiment Sheet 14

    Sink or Float 15

    What's in the Box? 16

Scientific Measurements 8

    Measure It–Volume 17

    Measure It–Length 18

    Latitude Finder 19

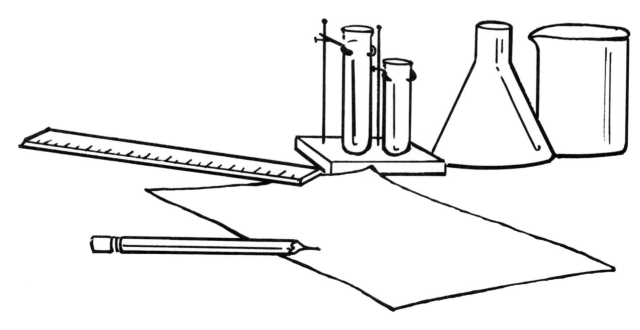

TEACHER NOTES

CLASSIFICATION

The process of classification is basic to the field of science. From the botanist or ornithologist categorizing a new plant or bird to the chemist's discovery of a new pharmaceutical drug, the process of organizing, ordering and classifying is in evidence.

To introduce the process of classification, gather materials for the first activity and arrange them so that the students can see all of the objects. Identify each object to be certain that each student understands what it is. If all students do not have a good view of the objects, list them on the chalkboard or a similarly visible spot. The object of this activity is to classify or organize in a group as many objects as possible. Encourage the use of objects in more than one group and classify/organize the items in as many different ways as possible.

Use the following example, or a similar example, to introduce the activity. Substitute similar objects for the ones listed if you do not have these exact items, as the categories are flexible. Make certain that you have items to fit the categories you devise.

The following objects, a pink pencil eraser, a pen, a pink flower, a seashell and a leaf, can be classified in these ways:

Color (pink)	Living	Use for writing	Inanimate Object
eraser	shell	eraser	all
flower	flower	pen	
	leaf		

These objects could also be categorized by size, smell and possibly by shape.

This same activity could be repeated using different objects and could function as an independent student activity or a science center to reinforce the concept of classification. For use as an independent activity, simply place the objects in a shoebox and provide the students with a space to spread them out and examine, sort and classify them.

The first activity in this section is a group activity or demonstration; the second and third are designed to be used as individual handouts.

Activity Page 10 Teacher Key — CLASSIFY IT

Place the following materials in a box for individual use or arrange them on a table for large-group viewing or use in a center. Substitute items if necessary but include a broad range of items and try to have more than one item for each category. The activity sheet which follows has space for six categories. Instruct the students to make additional categories on the back of the sheet or on notebook paper, or supply them with additional copies of the activity sheet.

<u>Suggested objects</u>: pen, pencil, crayon, scissors, comb, safety pin, different types of buttons, rocks, insects, seashells, flowers and seeds.

<u>Possible categories</u>: things with which to write/draw, things to be used for art, things to hold clothing together, living things, non-living things, plants, animals or parts of animals, edible and nonedible things.

Activity Page 11 Teacher Key — CLASSIFICATION TEASER

Four possible categories and the items in those categories are:

Things That Fly	Plants	Animals	Things That Have Shells
airplane	flower	turtle	turtle
bird	pear	clam	clam
butterfly	seed	snail	snail
		bird	
		butterfly	

Accept any other categories that are appropriate.

Activity Page 12 Teacher Key — CLASSIFY THE CRITTERS

Mammals	Amphibians	Insects	Reptiles
dog	frog	butterfly	snake
human being	salamander	dragonfly	turtle
cow		beetle	lizard
rabbit			alligator

Additional Classification Ideas

Teacher-Directed Activities

WEATHER WATCH

Have students sort pictures of clouds by cloud type (cumulus, stratus, or cirrus or further refinement into cumulonimbus, cirrostratus, nimbostratus, altocumulus, stratocumulus and altostratus). Additionally, the student should be able to describe the type of weather associated with each type of cloud.

Clip pictures of various clouds (at least two for each category) and a sample of each specific cloud type (titled for easy reference) and place in a manila envelope. The student sorts the pictures to identify each cloud type and matches it to the titled samples. Pictures are placed back in the envelope after completion of the activity. As an alternative, have students bring in photographs or pictures of clouds from magazines and arrange them on a poster or bulletin board under the correctly labeled sample clouds.

COLLECTIONS

Assemble and label (with common and scientific names) a collection of insects, shells, leaves or plants. For a plant collection, leaves and small plants can be pressed between the pages of a large telephone book or between sheets of paper under a large stack of books. To ensure durability, they can be ironed between two pieces of waxed paper or laminated. If you want students to collect rocks, have them collect as many different kinds of rocks as possible. They should then label each rock by rock type, and list on paper the geographic area in which each specimen was originally located.

To create a more in-depth study, have students identify the items in their collections by common and scientific names and list relevant comments explaining the difference between items. When developing a seashell collection the shells should be separated into univalves and bivalves. Fresh water and salt water shells should be distinguished and pertinent information listed for each member of the collection. For example, the murex, Murex trunculus (saltwater, univalve) was used by the ancient Persians to dye cloth purple.

THE SCIENTIFIC METHOD

Defining the scientific method for the students and then having them follow the method through an actual experiment allows students to understand how to solve a scientific problem and then record their methods and results in a manner that will enable them to retrace their work and check their conclusions. For this reason, the first step is to state and define the scientific method and make certain that each student understands the terms. The student activity sheet on page 13 is designed to give students an overview and introduce the FORMAL EXPERIMENT SHEET on page 14.

The FORMAL EXPERIMENT SHEET can be used for all student experiments. The SCIENTIFIC LAW section may not always be relevant. Activity sheets provided in this book can be used in addition to or in place of the experiment sheet. They provide a good framework for a science notebook. Students can complete response sheets on all experiments and file them in a loose-leaf binder.

Activity Page 13 Teacher Key

THE SCIENTIFIC METHOD

1. No. The hypothesis is a guess. The experiments prove whether the hypothesis is correct or incorrect.
2. If the experiments are not done carefully, the results will not be accurate. Inaccurate results do not prove anything.
3. No. The theory is unproven. Although it is based on observed fact and some experimentation, it may still not be correct.
4. It should be correct. The scientific law has been tested by experiments and found to be true in all known situations.

As an example of how to adapt the student activity sheets to the formal experiment format, the answer key to SINK OR FLOAT is incorporated with the FORMAL EXPERIMENT format on the following page.

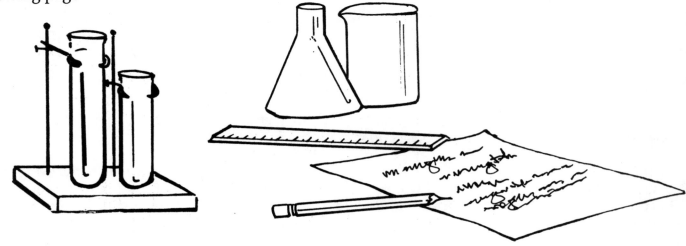

© Frank Schaffer Publications, Inc.   FS-10112 Hands-On Science Experiments

Activity Page 14 & 15 Teacher Key

THE SCIENTIFIC METHOD

SINK OR FLOAT

PROBLEM: What causes objects to sink or float in water?
FACTS: Some objects float in water; some objects sink.
    1. & 2. See Scientific Law and Hypothesis.

HYPOTHESIS: Students may suggest weight as a factor. Light objects float; heavy objects sink. Students may also suggest shape or air content (objects that are full of air float). Accept all reasonable suggestions and test them.

EXPERIMENT:
    Materials: pan of water, small jar with a lid, leaf, toothpick, dry paper, buttons made of different materials and of different shapes and sizes (include at least one button made of wood or leather), small plastic boat.

    Procedure:
    1. The empty, covered jar floats.
    2. The partially filled jar sinks a little.
    3. The full jar sinks.
    4. The leaf floats.
    5. The toothpick floats.
    6. The dry paper floats.
    7. Most metal and plastic buttons sink, unless they are extremely light and flat or hollow and airtight. Wood and leather buttons will float until they become thoroughly waterlogged.
    8. Most plastic boats float. If this one didn't, try to help the students determine why.

RESULTS:
1. An empty, covered jar floats.
2. A jar partially filled with water sinks a little.
3. A jar full of water sinks.
4. A leaf floats
5. A toothpick floats.
6. Dry paper floats. Wet paper sinks.
7. Most metal and plastic buttons sink unless they are extremely light and flat or hollow and airtight. Wood and leather buttons float until they become thoroughly waterlogged.
8. A plastic boat floats.

THEORY: Objects which contain air or air pockets float: empty jar, leaf, dry paper, wood, leather, boat-shaped items. Light weight objects float.

SCIENTIFIC LAW: Lighter (less dense) objects float. Heavier objects can float if the weight is distributed over a larger surface of water. The heavier the object, the more water surface it needs to cover. This includes water surface under and around the object and not just flat across the top of the water.

THINK:
    1. Attach balloons to it, change its shape, fill it with air . . . .
    2. They are wet, they do not contain air, they are the wrong shape . . . .

© Frank Schaffer Publications, Inc.

WHAT'S IN THE BOX?

WHAT'S IN THE BOX? is a good introductory activity for the beginning of a unit or the start of a new school year. It can be done as a whole-class activity, as a demonstration, or as a center activity with the students taking turns investigating the contents of the containers. Collect all materials and prepare the containers before class. Place them out of sight so that the students do not know the contents of the containers before they begin the activity. Before beginning this activity, review with the students five senses: sight, sound, smell, taste and touch. This is a question on the students' activity page 16 as well.

TEACHER PREPARATION FOR WHAT'S IN THE BOX?
MATERIALS: Four boxes, one clear container (or open dish), a magnet, a bell, a sliced onion, a comb, a nail and a half cup of either salt or sugar.

PROCEDURE:
1. Seal the bell in a box so that it can be heard if the box is shaken but cannot be felt or seen.
2. Seal the nail in a box as you did the bell in step one.
3. Seal an onion in a box, but punch small holes in one side of the lid so that the smell of the onion is easily detectable without opening the box.
4. Cut a hole in one box so that a student could fit his/her hand through the opening. The comb goes in this box. Make certain that the comb is not visible by covering the opening with a fabric or paper "door."
5. Place the salt or sugar in the container.
6. Place all the items on a table in the front of the room or in a similarly visible spot. Label each box with a number (the bell is number one, the nail is number two, the onion is three and the comb is four) and place a card with the number five on it by the salt or sugar.

Activity Page 16 Teacher Key

WHAT'S IN THE BOX?

1. Sight, hearing, taste, touch and smell are the five senses.
2. You can use instruments to help you identify unknown objects.

WHAT TO DO:
1. Box three–the onion can be identified by smell.
2. Box four–the comb can be identified by touch.
3. Box one–the bell can be identified by hearing.
4. Yes. Salt or Pepper.
5. No. It attracts a magnet and is made of iron.

Follow this experiment by discussing the following questions:
1. How do people who are blind or deaf adapt to compensate for the loss of one of their senses? How much more difficult is the loss of both sight and sound? (Have the students research the life of Helen Keller.)
2. Which sense is most necessary? Why? Here, consider that without touch, a person feels no pain. For example, the disease leprosy causes numbness and insensitivity to pain. Lepers can cut or burn themselves without being aware of it. Without the sense of taste, what might a person experience?

SCIENTIFIC MEASUREMENTS

The following chart should be available in student textbooks. If it is not, provide copies for the students.

METRIC CHART

Units of Length

1000	Millimeters	(mm)	=	1	Meter
100	Centimeters	(cm)	=	1	Meter
10	Decimeters	(dm)	=	1	Meter
1	Dekameter	(dkm)	=	10	Meters
1	Hectometer	(hm)	=	100	Meters
1	Kilometer	(km)	=	1000	Meters

Units of Volume

1000	Milliliters	(ml)	=	1	Liter
100	Centiliters	(cl)	=	1	Liter
10	Deciliters	(dl)	=	1	Liter
1	Dekaliter	(dkl)	=	10	Liters
1	Hectoliter	(hl)	=	100	Liters
1	Kiloliter	(kl)	=	1000	Liters

Units of Weight

1000	Milligrams	(mg)	=	1	Gram
100	Centigrams	(cg)	=	1	Gram
10	Decigrams	(dg)	=	1	Gram
1	Dekagram	(dkg)	=	10	Grams
1	Hectogram	(hg)	=	100	Grams
1	Kilogram	(kg)	=	1000	Grams

**Activity Page 17
Teacher Key**

MEASURE IT–VOLUME

BEFORE YOU START:
1. 16 cups in a gallon.
2. 8 pints in a gallon.
3. 10 deciliters in a liter.
4. 100 centiliters in a liter.
5. 1000 milliliters in a liter.

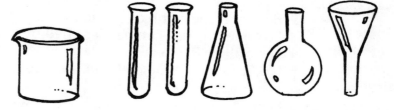

Which system of measurement do you think is easier? Why? (Accept any reasonable answer.)
Following the activity, discuss grams and pounds and the ease of using the metric system of weight. Compare how many ounces in a ton to how many grams in a kilogram. (16 ounces = 1 pound, 2000 pounds = 1 ton) Have students calculate percentages in each system. (67% of a pound versus 67% of a hectogram)

© Frank Schaffer Publications, Inc.

Activity Page 18 Teacher Key

MEASURE IT–LENGTH

1. 27 inches in 75% of one yard.
   75 centimeters in 75% of one meter.
2. 1584 feet in 30% of one mile.
   300 meters in 30% of one kilometer.
3. 352 yards in 20% of one mile.
   20 dekameters in 20% of one kilometer.

WHAT TO DO:
1. The meter stick is larger.
   An inch is larger than a centimeter.
2. 39.37 inches in one meter. (Accept any answer that is close.)
   A little less than three centimeters in one inch.
   25 millimeters in one inch.

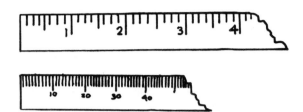

Activity Page 19 Teacher Key

LATITUDE FINDER

The latitude finder, a primitive sextant, can be made in class; however, it can only be used at night, since the North Star cannot be seen during the day. Using a sextant, students take turns sighting an object in the room and recording their findings. The readings may vary depending on the placement of the chosen object and the relationship of each student to that object, but this activity will give them experience using the instrument.

Since the latitude finder is primarily a "make-and-take" activity, there is not a teacher key for the questions. Have students report their findings the next day and check their readings with the latitude on a globe or map. Be certain that students know how to find the North Star before they take readings from the night sky. The latitude finder activity sheet (page 19), has a diagram to help students identify the North Star; however, most students are not very familiar with constellations and would profit from extra assistance. Have students practice identifying the major constellations from diagrams or individual constellations, and then locate the constellations on a large sky chart.

Name _____

CLASSIFY IT

In the boxes below, write the category of the reason for the items being placed in that list. Under the title, list each thing that belongs in that category.

Name_____

CLASSIFICATION TEASER

WHAT TO DO: Look carefully at the pictures in the middle of the page. Classify these items into four different categories. Create a title that explains the common element of the group in the box at the head of each list. Write the names of the items under that title.

© Frank Schaffer Publications, Inc.   FS-10112 Hands-On Science Experiments

Name_____

CLASSIFY THE CRITTERS

WHAT TO DO: Look carefully at the pictures in the middle of the page. Classify each item under the correct title in the lists below. In the blank spaces in each list, try to name at least one other animal that could be included in that class.

MAMMALS	AMPHIBIANS
_____	_____
_____	_____
_____	_____
_____	_____

INSECTS	REPTILES
_____	_____
_____	_____
_____	_____
_____	_____

Name_____

THE SCIENTIFIC METHOD

1. THE PROBLEM: Clearly state the question which you want to answer. Next, gather as many facts as you can about this problem.

2. THE HYPOTHESIS: The hypothesis is a reasonable solution to the problem. It is a guess, your idea of the possible answer to the problem. It may or may not be correct; that is what the experiment(s) will show.

3. THE EXPERIMENT: This is the way you test your hypothesis. If the results of the experiment don't agree with your hypothesis, you may have to change your hypothesis. You also need to be careful when you do the experiments and make certain that you are doing them correctly and recording the results accurately.

4. THEORY: After the experiments are finished, and the hypothesis seems correct, the scientist develops a theory which explains the facts and should predict the results of the experiments.

5. SCIENTIFIC LAW: This is the general statement which explains all of the facts and the results of the experiment(s). It explains why the experiments came out the way they did. Basically, a theory is accepted as a law after it has been proven to be correct and is found to be true in all situations.

DID YOU UNDERSTAND?

1. Will your hypothesis always be correct? Why or why?_____
   _____
   _____

2. Why do you need to be careful when you are doing experiments? _____
   _____
   _____

3. Will your theory always be correct? Why or why not? _____
   _____
   _____

4. Will the scientific law always be correct? Why or why? _____
   _____
   _____

Name_____

FORMAL EXPERIMENT SHEET

PROBLEM: _____
    Facts: _____
    _____
    _____

HYPOTHESIS: _____
_____

EXPERIMENT:
    Materials: _____
    _____
    _____

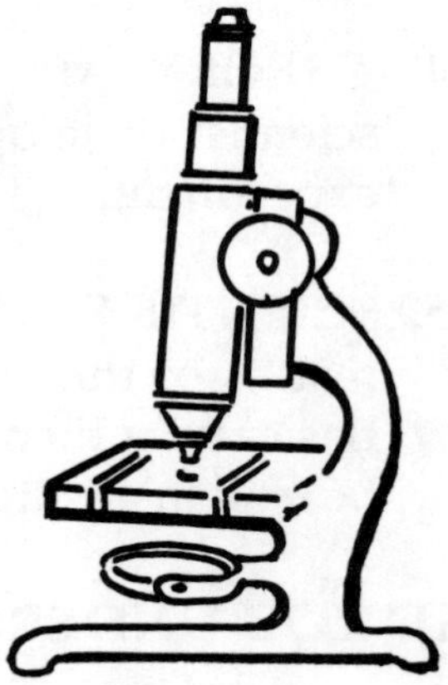

    Procedure: _____
    _____
    _____
    _____
    _____
    _____
    _____

RESULTS: _____
_____
_____
_____
_____
_____

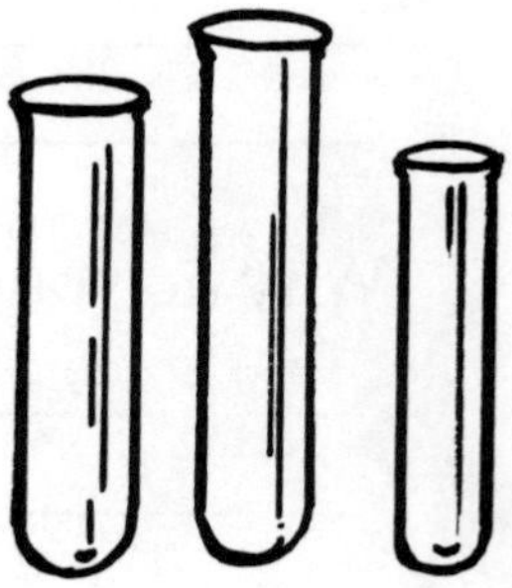

THEORY: _____
_____
_____

SCIENTIFIC LAW: _____
_____

© Frank Schaffer Publications, Inc.   FS-10112 Hands-On Science Experiments

SINK OR FLOAT

Name_____

PROBLEM: What causes objects to sink or float in water?

Facts: _____
_____

1. Why do you think some objects float in water? _____
2. Why do you think some objects sink? _____
_____

EXPERIMENT:
MATERIALS: A pan of water, a small jar with a cap, a leaf, a toothpick, a piece of dry paper, buttons made of different materials and of different shapes and sizes (at least one button should be made of wood or leather), a small plastic boat.

PROCEDURE:
1. Place the empty, covered jar in the pan of water.
   What happens? _____
2. Take the jar out of the pan, pour a little water in the jar, place the lid on the jar and place the jar back into the pan.
   What happens? _____
3. Take the jar out of the pan again, fill it up with water, place the lid on the jar and place it back into the pan.
   What happens? _____
4. Place the leaf on top of the water.
   What happens? _____
5. Place the toothpick on top of the water.
   What happens? _____
6. Place the paper on top of the water.
   What happens? _____
   Leave the paper in the water until it becomes soggy and completely wet.
   What happens? _____
7. Place the buttons on top of the water one by one. Tell what happens to each button separately. _____
   _____
8. Place the plastic boat on top of the water.
   What happens? _____

THINK: 1. How could you make a heavy object float? _____
   _____
   2. Why do some light objects sink? _____
   _____

Name_____

WHAT'S IN THE BOX?

1. What are your five senses? _____
   _____

2. What else could you use to help you identify something? _____
   _____
   _____

EXPERIMENT:
WHAT YOU NEED: Your five senses, a magnet

WHAT TO DO:
1. Examine each box and the container without touching them. Can you tell what is in any of the boxes? If so, how can you tell? _____
   _____

2. Reach inside the box with the opening (#4). Can you tell what is inside? If so, how can you tell? _____
   _____

3. Shake boxes #1, #2 and #3. Can you tell what is inside any of them? If so, how can you tell? _____
   _____

4. Taste the ingredients in container #5. Can you tell what is in the container by tasting? _____ What is it? _____

5. Rub a magnet against box #2. Can you tell what is in the box? _____
   What can you tell about what is in the box? _____
   _____

THINK: What other instruments could you use to find out what is in box #2 without opening the box? _____
_____

Name_____

MEASURING IT—VOLUME

Why do scientists prefer to use the metric system?

FACTS YOU NEED TO KNOW: The U.S. Customary system of measuring volume uses cups, pints, quarts and gallons.

2 cups   = 1 pint
2 pints = 1 quart
4 quarts = 1 gallon

The metric system of measuring volume uses milliliters, centiliters, deciliters, dekaliters, hectoliters and kiloliters.

1000 Milliliters (ml) = 1 Liter
100 Centiliters (cl) = 1 Liter
10   Deciliters (dl   = 1 Liter

BEFORE YOU START:
1. How many cups of water are there in a gallon of water? _____

2. How many pints of water are there in a gallon of water? _____

3. How many deciliters of water are there in a liter of water? _____

4. How many centiliters of water are there in a liter of water? _____

5. How many milliliters of water are there in a liter of water? _____

WHAT YOU NEED: A one-cup measuring cup, a pint container, a quart container, a gallon container, a liter container and a measuring cup/flask with deciliter, centiliter and milliliter markings on it.

WHAT TO DO: Check your answers to the questions above by measuring.

Which system of measurement do you think is easier? Why?

MEASURING IT–LENGTH

Why do scientists prefer to use metric measurements for length?

FACTS YOU NEED TO KNOW: The U.S. Customary system of measuring length uses inches, feet, yards and miles.

12 inches = 1 foot
3 feet = 1 yard
5280 feet = 1 mile

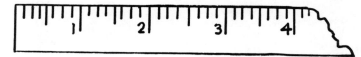

The metric system of measuring length uses millimeters, centimeters, decimeters, dekameters, hectometers and kilometers.

1000	millimeters	(mm)	=	1	meter
100	centimeters	(cm)	=	1	meter
10	decimeters	(dm)	=	1	meter
1	dekameter	(dkm)	=	10	meters
1	hectometer	(hm)	=	100	meters
1	kilometer	(km)	=	1000	meters

1. How many inches are there in 75% of one yard? _____
   How many centimeters are there in 75% of one meter? _____
2. How many feet are there in 30% of one mile? _____
   How many meters are there in 30% of one kilometer? _____
3. How many yards are there in 20% of one mile? _____
   How many dekameters are there in 20% of one kilometer? _____

WHAT YOU NEED: A meter stick and a yardstick.

WHAT TO DO:
1. Place the sticks side by side and compare them. Which is larger, the meter stick or the yardstick? _____
   Which is larger, a centimeter or an inch? _____
2. Measure:
   How many inches in one meter? _____
   How many centimeters in one inch? _____
   How many millimeters in one inch? _____

LATITUDE FINDER

WHAT YOU NEED: A sheet of construction paper, a protractor, tape, string and a weight (a fishing weight, a key or a large pencil eraser will work).

WHAT TO DO:
1. Roll the paper into a tube 1/2 inch in diameter and place tape around it to hold it together.

2. Tape the protractor to the paper tube with the flat part against the tube and the curved edge curving away from the tube as shown.

3. Tie your weight on one end of the string. Tape the other end to the paper tube and the protractor at the midpoint of the straight edge of the protractor.

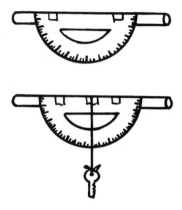

HOW TO USE YOUR LATITUDE FINDER: Take your latitude finder outside on a clear night when you can see the stars. Looking through the paper tube, find the North Star. Trying not to move your latitude finder, use one hand and carefully hold the string against the bottom, curved side of the protractor. Holding the string in place, take the latitude finder from your eye and look to see what number the string touches. Your protractor has two sets of numbers. Use the lower number. This is your latitude.

FINDING THE NORTH STAR: The constellation Ursa Major, which we usually call the Big Dipper, points to the North Star (Polaris). The North Star is the tip of the handle of the Little Dipper (Ursa Minor).

CHAPTER TWO

THE LIVING WORLD–BIOLOGY

Teacher Notes 21
Biological Organization–Classifying the Living World 21
    Classifying Plants–Seeds..31
    Classifying Plants–Leaves 32
    Plant or Animal 33
    Classifying Animals 34
Microorganisms 22
    Classifying Microorganisms 35
    Diffusion 36
    Droplet Infections 37
Plants 23
    Cold Weather Seeds 38
    Traveling Seeds..39
    Tropisms 40
Insects 25
    Insect Life Cycles 42
    Useful Insects and Pests 43
Marine Life 25
    Fishy Facts 44
    Coral Reef Wall 45
    The Coral Reef..46
Mammals 26
    Types of Mammals I 47
    Types of Mammals II 48
The Human Body 27
    The Skeleton 49
    Forensic Specialist 50
Prehistoric Plants and Animals 29
    Dinosaur Celebrities..51
    Making Fossils 52
Genetics...30
    Genetics 53

TEACHER NOTES

BIOLOGICAL ORGANIZATION—CLASSIFYING THE LIVING WORLD

Additional activities that apply to this section can be found in the first section of Chapter One. The following student activity sheets require no additional materials; however, the activities could be extended by providing actual specimens for the students to classify.

Activity Page 31 Teacher Key

CLASSIFYING PLANTS—SEEDS

MONOCOTS	DICOTS
corn	peach
date palm	peanut
rice	lima bean
grass	pea
	maple

CLASSIFYING PLANTS—SEEDS

CLASSIFYING PLANTS—LEAVES (page 32) is simply a guide sheet to be used in classifying actual leaves. Have students take the sheet home or on a field trip to an area where gathering leaf specimens is allowed and collect an example of each category. If actual samples of leaves cannot be collected, take a nature walk and identify as many different trees and plants as possible. Students should record the name of each specimen identified and where it was found. They can also write a brief description and draw a sketch. The categories overlap as stated on the student sheet; therefore, one leaf could be placed in as many as four categories.

As a followup activity, allow students to find facts by using the library, *National Geographic Magazine* or an encyclopedia.

Activity Page 33 Teacher Key

PLANT OR ANIMAL

Common Classification: I.

PLANT	ANIMAL
pine tree	Gila monster
tomato	wolf spider
Venus flytrap	sponge
	coral

II. Accept any true facts about these plants and animals.

Activity Page 34 Teacher Key

CLASSIFYING ANIMALS

ARTHROPODA	CHORDATA	MOLLUSCA
spider	fish	octopus
crab	frog	clam
dragonfly	bird	snail
centipede	human being	

MICROORGANISMS

One of the best ways to introduce students to the study of microorganisms is to have them view the organisms directly through a microscope. Place a drop of pond water on a slide. Allow students to observe the slide using a microscope. Have them draw what they see and then label their observations. Use the first student activity sheet, CLASSIFYING MICROORGANISMS, to help them identify their findings and then classify the microorganisms.

Activity Page 35 Teacher Key

CLASSIFYING MICROORGANISMS

PROTOZOA	ALGAE
amoeba	chlamydomonas
paramecium	desmids
stentor	spirogyra
euglena	rhodophyta—Lemanae
volvox	vaucheria
	dinobryon

Activity Page 36 Teacher Key

DIFFUSION

Demonstrate osmosis using a raisin or prune. The skin of the dried fruit is a semipermeable membrane. Place the fruit in a glass of hot water and leave it untouched overnight. The membrane (skin) will allow water molecules to pass through it, and the dried fruit will become plump and water-filled.

THINK: Sugar in coffee or tea; salt in soup or water; perfume, onion or any smell in the air; salt and minerals and pollutants in the oceans, rivers and lakes; chocolate syrup in milk; smoke in the air; soap or detergent in water, etc.

DISEASES CAUSED BY MICROORGANISMS

One method of transmitting disease is discussed in this section: droplet infections. Contact infections, contamination infections, wound infections, vector-carried infections, immune carriers, and methods for preventing the spread of disease should all be additional topics for discussion.

Activity Page 37
Teacher Key

DROPLET INFECTIONS

6. and 7. Answers should reflect the student's observations.

THINK:
1. A sneeze usually has more force than a cough, so the student should guess as far or farther than a cough.
2. The pathogens would land on the nearby surfaces.
3. Cover mouth and nose when coughing or sneezing, wash hands, dispose of used tissue and do not share food or drinking containers.

For a follow-up activity, have students research the diseases of the third world countries, where clean water supplies and medical assistance are less available. In addition, discuss immunization and the effect that a strict immunization program from the 1950s to the 1970s had on polio. Note that polio is now reappearing as some parents neglect to have their children vaccinated for a disease that they think is no longer a threat. How can we stop the spread of diseases, particularly life-threatening diseases? Discuss possible methods of prevention.
Classification of seeds and plants is treated in section one of this chapter.

PLANTS

Activity Page 38
Teacher Key

COLD WEATHER SEEDS

WHAT HAPPENED:
Students should record their observations.

THINK:
1. The answer should be the same as the answers to questions three and five in the WHAT HAPPENED section.
2. The answer should be the same as the answer to question two in the WHAT HAPPENED section.

Activity Page 39 Teacher Key

TRAVELING SEEDS

1. Dandelion–airborne
2. Beggar's lice–attaches to animals and people
3. Cherry–carried by animals or people
4. Coconuts–water floated, carried by animals or people
5. Maple–airborne
6. Grass–airborne or carried by animals or people

THINK: Dandelions were brought to America deliberately by colonists. They were used for salad greens and dandelion wine.

Activity Page 40-41 Teacher Key

TROPISMS

NOTE: The following three activities are all related. They are grouped on two student activity pages to be presented simultaneously.

GEOTROPISM
4. The answer should reflect student observation.
5. The answer should reflect student observation.

THINK: 1. Plants tend to grow upward from the earth.
2. Gladiola bulbs sprout upward from the earth.
3. Acorns are similar to gladiola bulbs. (Gladiola bulbs are easier to observe because they have such a defined "top" and "bottom," but other seeds and nuts react similarly.)

PHOTOTROPISM
1. The answer should reflect student observation.
2. The answer should reflect student observation.

THINK: Plants grow in the direction of the sunlight.

HYDROTROPISM
3. The answer should reflect student observation.
4. The answer should reflect student observation.

THINK: 1. Plant roots grow in the direction of water.
2. If the plants are soaked, the roots tend to extend down into the soil. If the plants are given a light sprinkle, the roots tend to stay on top of the soil and cannot weather severe drought.

INSECTS

Insect collections are a fascinating way for students to learn about the world of insects. Students may mount their specimens on posterboard or in large boxes and label them. They could also make small booklets or do reports of interesting facts about their specimens. Make certain they know that spiders are not insects.

Activity Page 42 Teacher Key

INSECT LIFE CYCLES

COMPLETE METAMORPHOSIS: butterfly, mosquito
INCOMPLETE METAMORPHOSIS: praying mantis, dragonfly, grasshopper
NO METAMORPHOSIS: silverfish, lice

Activity Page 43 Teacher Key

USEFUL INSECTS AND PESTS

Quick Quiz: The praying mantis, ladybug and bee are useful.
The mosquito and housefly are pests.

1. Praying mantises eat other bug pests.
2. Mosquitoes carry many diseases, such as malaria, and spread the diseases through their stings.
3. Ladybugs eat aphids which can destroy plants.
4. Bees make honey and pollinate crops.
5. Houseflies carry many diseases, such as salmonella, and transmit them to people by crawling on uncovered food.

Mantis egg cases and ladybugs can be purchased through gardening magazines and some shops. If you purchase an egg case, enclose it in a container with pinholes for ventilation. Mantis nymphs are very lively. They would be an excellent addition to a terrarium and, by using dirt and wild plants, you will provide them with their first dinner. Remember to cover the terrarium as mantises move quickly. Hatch the eggs in the spring when you can release the mantises or ladybugs outside the classroom.

© Frank Schaffer Publications, Inc.   FS-10112 Hands-On Science Experiments

MARINE LIFE

Marine life is so varied and extensive that students have myriad choices for topics to research. In addition, a trip to an aquarium, or tropical fish store, can spark enthusiasm for this wonderful world. Books, films, videos and television programs on sea life will also help to make marine life forms more familiar.

Activity Page 44 Teacher Key — **FISHY FACTS**

1. Each piece of seastar grew into a whole new seastar, creating many more seastars, not fewer, as the fishermen had intended.
2. The flounder's shape and color provide camouflage to protect it from its enemies.
3. The sponge acts as camouflage for the crab.
4. The blenny gets an easy meal.

Activity Page 45 Teacher Key — **CORAL REEF WALL**

All of the answers should reflect actual student observations.
3. Gentle waves should move very little sand.
4. Larger waves should move some sand, but the sand should not wash beyond the gravel barrier.
5. The gentle waves move more sand than before.
6. The large waves erode the sand.

THINK: Coral reefs protect the islands and beaches and keep them from being eroded and washed away.

Activity Page 46 Teacher Key — **THE CORAL REEF**

1. stony coral
2. plankton
3. algae
4. sponges
5. microscopic animals
6. spiny sea urchins
7. seastar
8. sea cucumber
9. clam
10. oyster

Extend the CORAL REEF activity sheet by researching the relationship of the organisms listed above. For example, the seastars eat the coral and the coral also feeds on the eggs of the seastar.

MAMMALS

The class Mammalia contains a great variety of animals, many of which are familiar to the students. Zoo trips and daily observation of wildlife such as squirrels and domestic animals can enrich the students' experience. Have them observe animals and note their behavior, eating habits, habitats and relationships with other animals. What animals share habitat space and what animals cannot share the same space? Why or why not? What animals are territorial? Have the students keep notebooks and record their observations daily.

> **Activity Page 47
> Teacher Key**

TYPES OF MAMMALS I

MONOTREMES	MARSUPIALS	RODENTS
duckbilled platypus	opossum	mouse
spiny anteater	kangaroo	squirrel
	koala	rat
	Tasmanian devil	porcupine

> **Activity Page 48
> Teacher Key**

TYPES OF MAMMALS II

CARNIVORES	AQUATIC MAMMALS	PRIMATES
tiger	whale	baboon
seal	dolphin	ape
raccoon	porpoise	marmoset
bear	sea lion	

Enrichment activities and topics which may be pursued: research the platypus, Tasmanian devil, spiny anteater and marmoset. Listen to recordings of whale songs and wolf howls and discuss the patterns of sounds and the intelligence, social development and group relationships of these animals. Research and compare and contrast the hunting patterns of wolves, tigers and bears. Research the relationship humans have with whales, dolphins, porpoises and seals.

THE HUMAN BODY

The human body provides a tremendous opportunity for student research, demonstrations and activities. Discuss voluntary and involuntary actions with the students and have them demonstrate the response of the knee to tapping and the eye to blinking when an object comes near the eye. Invite a doctor or nurse to demonstrate, explain and discuss cholesterol levels, blood types and blood screening tests.

> **Activity Page 49
> Teacher Key**

THE SKELETON

1. cranium	6. scapula	11. carpals	16. pubis	21. fibula
2. maxilla	7. ribs	12. metacarpals	17. ischium	22. tarsals
3. mandible	8. humerus	13. phalanges	18. femur	23. metatarsals
4. clavicle	9. radius	14. sacrum	19. patella	24. calcaneus
5. sternum	10. ulna	15. ilium	20. tibia	25. phalanges

THINK: The skeleton supports the muscles and the body.
The diagram lists the bones of the body in the same order as given above for the activity page. Note that most of the bones are multiples.

© Frank Schaffer Publications, Inc.

BONES OF THE BODY

Cranium (1); Maxilla (2); Mandible (1); Clavicle (2); Sternum (1); Scapula (2); Ribs (24): 7 pair true ribs (true ribs attach directly to the sternum), 3 pair false ribs (do not attach directly to the sternum), 2 pair floating ribs (do not attach to sternum); Humerus (2); Radius (2); Ulna (2); Carpals (16;) Metacarpals (10); Phalanges (28): 3 in each finger, 2 in each thumb; Sacrum: consists of five fused vertebrae; Ilium (2); Pubis (2); Ischium (2); Femur (2); Patella (2); Tibia (2); Fibula (2); Tarsals (14); Metatarsals (10); Calcaneus (2); Phalanges (28): 2 in each big toe and 3 in every other toe.

Activity Page 50 Teacher Key

FORENSIC SPECIALIST

FORENSIC SPECIALIST is a game of "count the bones." Before handing out this sheet, review the bones of the body and note which bones are "one of a kind" and which ones come in pairs or multiples. After completion, discuss the solutions to each "crime" and the reason why each answer is the most likely choice.

1. These may be the bones of Mr. and Mrs. Ponsonby Smythe. (Three pelvic bone structures indicate 1 1/2 bodies.)
2. Yes. These could be the bones of Notso Bryte. (Half of a rib cage, one sternum and one humerus suggest one body.)
3. These are more likely to be the bones of Hy Wave. (A single skull and one complete hand indicate one body.)
4. These may be the bones of Sy and Amy Fuzzifer. Two sternums, 1 1/2 rib cages and three patellas suggest two bodies.)

In addition to the study of parts of the body and the functions of different parts of the body and body systems, there are several other interesting experiments that are simple and fun.

1. To demonstrate the center of gravity of the human body, place a stool three inches from the wall. Have the student bend over the stool and place his or her head against the wall. The legs and upper body should form a right angle. Next, instruct the student to pick up the stool and then straighten up without touching anything but the stool. Females and young boys can usually perform this task because their body weight and center of gravity are in the middle or lower part of their torso. Muscular males have their body weight and center of gravity in their upper torso and, consequently, cannot perform this task.

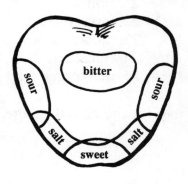

2. Test the sense of taste. Different parts of the tongue detect different flavors. Have samples of sugar, salt and something sour such as lemon and something bitter such as the lining of a pecan shell. Dip a damp cotton swab in the flavor and touch it to the different parts of the tongue.

3. Test the sense of touch with a pencil or pen that has flat sides. Touch the flat sides to a fingertip and feel the ridges. Touch it to the forearm and the ridges are not detectable. Touch it to the top of the nose and the ridges are not detectable. Touch it to the lips and feel the ridges. Touch it to the earlobes and the ridges are not detectable.

4. Test the sense of touch again with an ice cube. The earlobes and forearms are much more sensitive to cold than lips, nose and fingertips are.

5. Test peripheral vision. Have the subject sit in a chair and stare straight ahead. Stand behind the subject and extend hands out to either side of the subject just out of the subject's range of vision. Slowly bring hands together toward the front of the subject and have the subject report when he or she can see the movement of the hands.

6. Test hearing. Test to see whether high tones or low tones are easier to hear. (This will vary with the subject.) Also test the ability to determine the point of origin of a sound. Have the subject close his or her eyes and identify the direction from which a sound originated.

7. Test the effect of exercise on heart rate. Take the subject's pulse. Have the subject run in place for sixty seconds. Take the subject's pulse again. Repeat this with a new subject only have the second subject run for two full minutes. Is the difference greater? Have students test themselves and record the different pulse rates.

PREHISTORIC PLANTS AND ANIMALS

As an introduction or a summary to the topic of prehistoric plants and animals, discuss the plants and animals presently on the earth which bear a close resemblance to the plants and animals of prehistory. Note the resemblance between the ferns, coelenterates, turtles, crocodiles, alligators and sharks and their prehistoric counterparts. For example, blue-green algae are found in many places today and were one of the dominant organisms of the late Precambrian and early Paleozoic periods. The fossils are called stromolites and are composed of layered mounds of calcium carbonate and sediment. Stromolites are even being formed in a few places today, notably in Shark Bay off Western Australia.

In addition, discuss how theories change. Dinosaurs have been regarded as slow moving, cold-blooded and slow witted until recently. New findings have caused some scientists to challenge this theory and suggest that many were able to move quickly and were warm-blooded. Some may have been quite intelligent also.

Activity Page 51 Teacher Key

DINOSAUR CELEBRITIES

1. Allosaurus
2. Ankylosaurus
3. Triceratops
4. Apatosaurus
5. Stegosaurus
6. Iguanodon
7. Anatosaurus
8. Tyrannosaurus Rex

Carnivores: Allosaurus, Tyrannosaurus Rex, Iguanodon

Herbivores: Triceratops, Apatosaurus, Stegosaurus, Ankylosaurus, Anatosaurus

Activity Page 52 Teacher Key

MAKING FOSSILS

THINK: The animals on the bottom died first. Their bodies were covered with sediment, then other animals died and their bodies were covered with sediment. The process repeated, building up layer upon layer of skeletons and soil.

GENETICS

Before starting the following activity, be sure that students understand the concept of dominant and recessive genes and that the capital letter stands for the dominant gene and that the small letter stands for the recessive gene.

Activity Page 53 Teacher Key

GENETICS

1. Dark
2. Curly
3. Brown, hazel or green
4. Light, curly hair and blue or grey eyes
5. Dark, curly hair and brown, hazel or green eyes
6. Light, straight hair and brown, hazel or green eyes
7. Dark, straight hair and blue or grey eyes

Additional activities: Research genetic links to disease. Discuss the advances made in gene research and its application to criminal investigations. Compare photographs of actual genes and compare chromosomal composition.

CLASSIFYING PLANTS–SEEDS

Angiosperms, or flowering plants, are classified as either monocots or dicots. To decide whether a plant is a monocot or dicot, look at the seed to see how many cotyledons it has. The cotyledon is the part of the seed that stores food to feed the young plant as it grows. Dicot seeds will split easily into two equal parts. Monocots will not split.

WHAT TO DO: Look carefully at the seed pictures below. Decide whether they are monocots or dicots and then write the name of the seed in the list where it belongs.

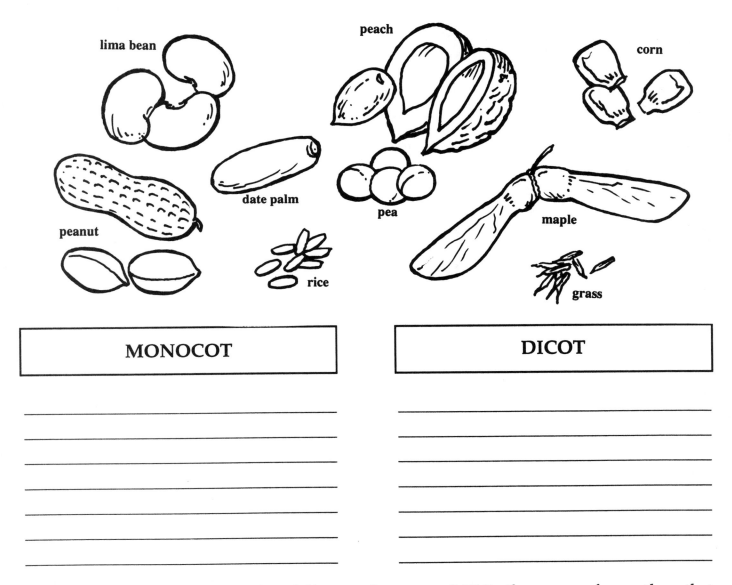

MONOCOT	DICOT
_____	_____
_____	_____
_____	_____
_____	_____
_____	_____
_____	_____

Do you know some more examples of dicots and monocots? Write the names of any others that you know in the extra blanks.

CLASSIFYING PLANTS—LEAVES

WHAT TO DO: Look at the chart below. Find an actual leaf for as many categories as possible. It may be possible to use the same leaf for two or three different categories. Press the leaves by putting them between two pieces of paper and then laying the papers on a flat surface under a heavy book or stack of books. When the leaves are flat and dry, glue them to a piece of notebook paper. Write the category or categories that fit each leaf beneath it.

MOSAIC. The way the leaves are arranged on the stem is called the leaf mosaic. The mosaic is either:

    Spiral   Alternate   Opposite   Whorled

MARGINS. The edges of the leaves are called margins. Margins are either:

   Entire   Undulate   Serrate   Dentate

SHAPES. Leaf shapes are either:

  Linear   Chordate   Deltoid   Lobed   Circular

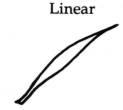    

VENATION. The leaf venation, or the way the veins of the leaf are arranged, is either:

Parallel   Simple   Pinnately   Bipinnately   Simple   Palmately
        Pinnate Compound   Compound   Palmate Compound

Name_____

PLANT OR ANIMAL

WHAT TO DO: Look at the plants and animals pictured below. Classify each one in the category in which you think it belongs. USE PENCIL.

PLANT	ANIMAL
_____	_____
_____	_____
_____	_____
_____	_____

Now, check your classification in an encyclopedia or guidebook. If your answer is not correct, change it.

Next, find one interesting fact about each of the following plants or animals and write it in the space following the name of the item.

1. Venus flytrap _____
   _____

2. Gila monster _____
   _____

3. Wolf _____
   _____

4. Sponge _____
   _____

5. Coral _____
   _____

MORE INTERESTING FACTS: Tomatoes and potatoes are from the same family of plants and they can cross pollinate. When they do, the potato plant produces tomatoes on top and potatoes on the bottom!

Carnivorous plants grow in soil that is nitrogen deficient. They must "catch" their dinner to provide enough nitrogen. One Malaysian pitcher plant is large enough to entrap and digest birds and small rodents.

© Frank Schaffer Publications, Inc.   FS-10112 Hands-On Science Experiments

Name_____

CLASSIFYING ANIMALS

WHAT TO DO: Look carefully at the pictures below. Write the name of the animal in the proper category. The categories given are three of the phyla for the kingdom Animalia.

ARTHROPODA exoskeleton, jointed legs	CHORDATA have backbones	MOLLUSCA soft bodies and a hard shell
_____	_____	_____
_____	_____	_____
_____	_____	_____
_____	_____	_____

© Frank Schaffer Publications, Inc.   34   FS-10112 Hands-On Science Experiments

Name_____

CLASSIFYING MICROORGANISMS

Look at the pictures of the microorganisms below. Some are protozoans, a term that means "first animals." Most protozoans can move. The others are algae, primitive plants. Write the name of each under the category to which it belongs.

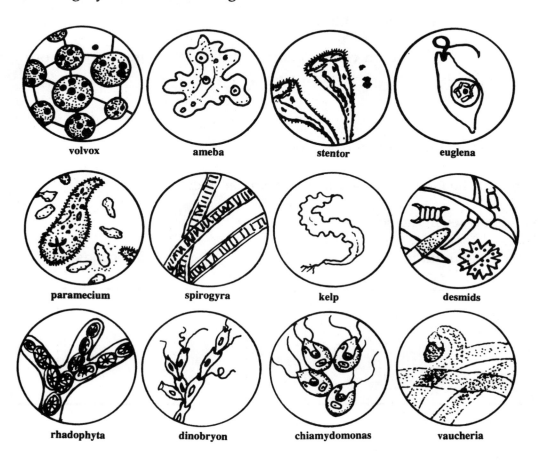

PROTOZOA	ALGAE

Name_____

DIFFUSION

Microorganisms do not have circulatory systems to move food molecules around in their bodies. Instead, they rely on the process of diffusion to spread the molecules around.

DIFFUSION: The movement of molecules from a place of high concentration (many molecules) to a place of low concentration (few molecules).

WHAT YOU NEED: A half gallon clear container full of tap water, food coloring and a 1/4 teaspoon measuring spoon.

WHAT TO DO:

1. Pour food coloring into the 1/4 teaspoon measuring spoon.

2. Carefully pour the food coloring into the container of clear water.

3. Draw what you see in the boxes below.

the instant you put food coloring into the container.

15 seconds later.

10 minutes later.

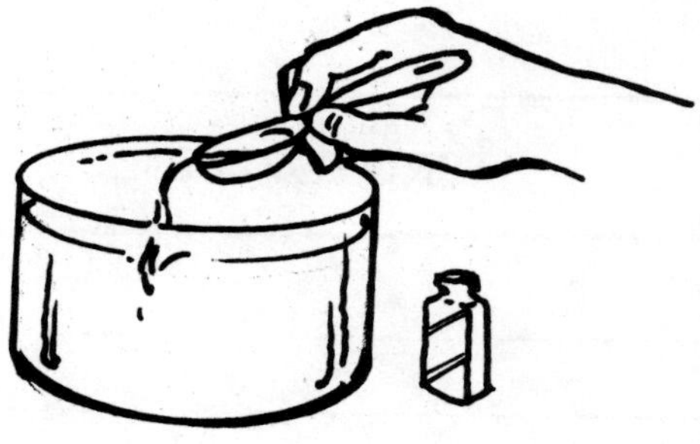

THINK: What are some everyday examples of diffusion? _____
_____

© Frank Schaffer Publications, Inc. FS-10112 Hands-On Science Experiments

Name_____

DROPLET INFECTIONS

DROPLET INFECTIONS: Many diseases are airborne and spread in water droplets that people sneeze or cough into the air. The pathogens that cause these diseases can't be seen with the naked eye. How far do these pathogens go when a person coughs or sneezes?

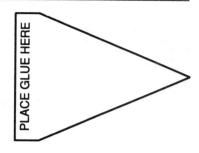

WHAT YOU NEED: Modeling clay, 10 toothpicks, a piece of notebook paper, scissors, glue and a pen or pencil.

WHAT TO DO:

1. Cut 10 triangles out of the paper. Use the pattern at the top of the page.

2. Number the triangles from 1 to 10.

3. Glue one triangle to each of the 10 toothpicks to make 10 pennants.

4. Form ten 1/2-inch balls from the modeling clay. Anchor a pennant in each ball of clay.

5. Place the pennants on a table one inch apart and standing in a straight line.

6. Have a healthy person put his or her mouth one inch from the first pennant and cough. How many pennants move? _____

7. Repeat step six. This time have the person blow at the first pennant. How many pennants move? _____

THINK:

1. How far do you think the pathogens would be carried if a sick person sneezed? _____

2. What would happen to those pathogens in the air after he or she sneezed? _____

3. How could a sick person prevent or lessen the spread of the pathogens? _____

Name_____

COLD WEATHER SEEDS

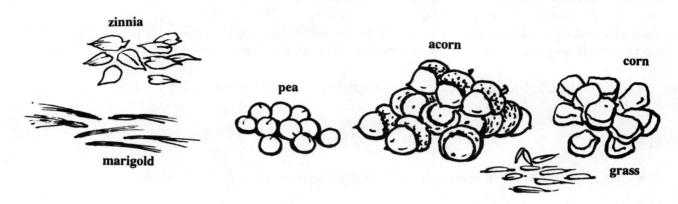

Why are seeds planted at different times of the year?

WHAT YOU NEED: 10 zinnia seeds, 10 grass seeds, 10 acorns, 10 marigold seeds, 10 pea seeds, 10 corn seeds, 12 containers, potting soil and water.

WHAT TO DO:
1. Label two containers for each kind of seed. Label one container of each kind of seed *warm* and one container *cold*.
2. Place potting soil in each container and plant five seeds of the same kind of plant in each container. Make sure the type of seed matches the label on the container.
3. Water the potting soil enough to dampen it but not enough to make it soggy.
4. Place the containers labeled *cold* in the refrigerator (not the freezer). Place the containers labeled *warm* in a warm place in the room. Keep both containers moist and out of direct sunlight.
5. Observe the containers daily for three weeks and record results.
6. After the containers sprout, put one container of each kind of sprouted seeds in a freezer for six hours.

WHAT HAPPENED:
1. Which container(s) sprouted first? _____
2. Which of the warm containers sprouted first? _____
3. Which of the cold containers sprouted first? _____
4. Which containers did not sprout? _____
5. Which containers survived the freezer? _____

THINK:
1. Which seeds would be good to plant in early spring or cool weather? _____
2. Which seeds should be planted in warm or hot weather? _____

Name_____

TRAVELING SEEDS

How do plants spread from one place to a place a long distance away?

WHAT TO DO: Look at the seeds pictured below. Beside each one, tell how you think they travel or are moved.

1. Dandelion   _____
_____

2. Beggar's lice  _____
_____

3. Cherry _____
_____

4. Coconut  _____
_____

5. Maple _____
_____

6. Grass _____
_____

THINK: How do you think some seeds crossed oceans? The dandelion was a popular plant in Europe but unheard of in America when the colonists first arrived. How do you think it got here?

_____
_____

Name_____

TROPISMS

Plants react to their environment by bending. This bending movement is called *tropism*. It is a response to one of three things: gravity, light or water.

GEOTROPISM: The way a plant responds to gravity.

WHAT YOU NEED: Four containers, potting soil, water, four bean seeds and two gladiola bulbs.

WHAT TO DO:
1. Fill each container with potting soil. Label two containers *Bean*. Label the third container *Gladiola Up* and the fourth container *Gladiola Down*.
2. Plant a bean in each bean container, water them and place the containers in a warm, sunny spot.
3. Plant a gladiola bulb in container three, right side up. Plant a gladiola bulb in container four, upside down. Check the diagram at the top of the page to see which side of the gladiola bulb is which. Water both containers and place them in a warm, sunny spot.
4. When the beans sprout, leave one container standing upright and turn the other on its side. What happens?_____
_____
5. When the gladiola bulbs sprout, dig them up. What has happened? _____
_____

THINK:
1. How does gravity affect plants? _____
_____

2. How does gravity affect gladiola bulbs? _____
_____

3. How do you think acorns would respond to gravity if you planted them? _____
_____

© Frank Schaffer Publications, Inc.   FS-10112 Hands-On Science Experiments

Name_____

PHOTOTROPISM: The way a plant responds to light.

WHAT YOU NEED: Two bean seeds, one container, potting soil and water.

WHAT TO DO:
1. Place potting soil in the container. Plant the bean seeds in the potting soil, water them and place the container in a window where it receives direct sunlight from the side and not from an overhead light. How do the beans grow? _____
_____

2. Turn the plants around so that they point away from the source of the sunlight. What happens to the bean plants? _____
_____

THINK: How do plants react to sunlight? _____
_____

HYDROTROPISM: The way a plant responds to water.

WHAT YOU NEED: A glass jar, a rubber band, a six-inch square of cheesecloth or netting, four bean seeds, cotton balls and water.

WHAT TO DO:
1. Fill the jar with wet cotton balls.

2. Cover the jar with the netting or cheesecloth and secure the netting with the rubber band.

3. Place the bean seeds on top of the netting and cover them with wet cotton balls. Place the whole jar in a warm, sunny place. Keep it wet. After five days, check the bean seeds. What has happened? _____
_____

4. After the beans have started growing and have roots, carefully remove the netting cover and empty the jar. Replace the netting and the bean seeds. Keep the cotton on top of the seeds wet. What happens?

THINK:
1. How do plants react to water? _____
_____

2. Why would it be better to give crops one good weekly soaking than a daily light sprinkle of water in dry weather? _____

© Frank Schaffer Publications, Inc. 41 FS-10112 Hands-On Science Experiments

Name_____

INSECT LIFE CYCLES

Insects go through different stages in their life cycles and may take different forms. The process of these stages and forms is called *metamorphosis*. Some insects do not go through metamorphosis; the young are simply small adults.

Incomplete Metamorphosis	**Complete Metamorphosis**
egg	egg
nymph	larva
adult	pupa
	adult

WHAT TO DO: Look at the insects and life cycles pictured below. Under each insect, write whether it has no metamorphosis, incomplete metamorphosis, or complete metamorphosis.

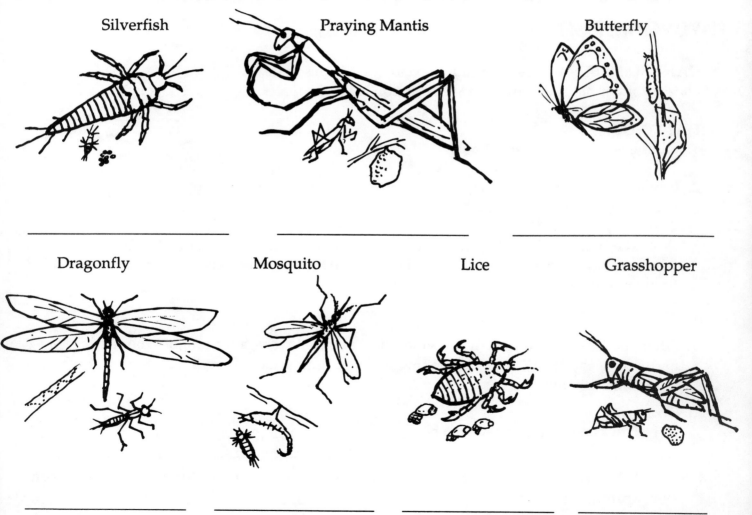

Silverfish   Praying Mantis   Butterfly

Dragonfly   Mosquito   Lice   Grasshopper

© Frank Schaffer Publications, Inc.   FS-10112 Hands-On Science Experiments

Name_____

USEFUL INSECTS AND PESTS

Quick Quiz: How well do you know your insects? Label the insects below, U for useful or P for pest.

Praying Mantis   Mosquito   Ladybug   Bee   Housefly

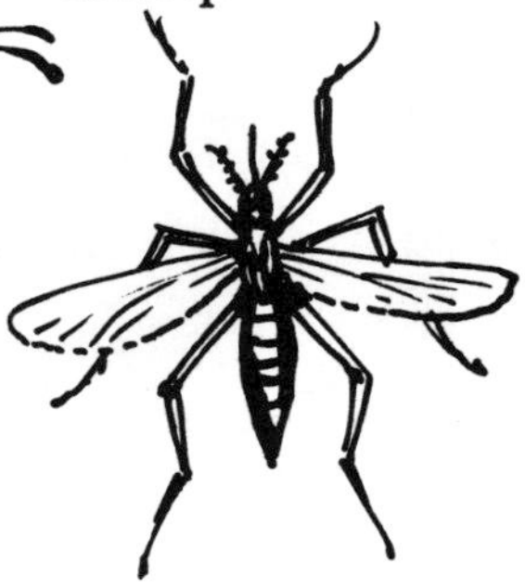

____   ____   ____   ____   ____

Now, explain why each insect is harmful or useful. If you don't know, research the insect in an insect guide book or encyclopedia.

1. Praying Mantis _____
   _____

2. Mosquito _____
   _____

3. Ladybug_____
   _____

4. Bee_____
   _____

5. Housefly_____
   _____

INTERESTING FACTS: Some "useful" insects turn out to be pests. The gypsy moth was brought to the United States intentionally by a man named Leopold Trouvelot. He wanted to produce silk and brought several different kinds of silkworms to our country. He finally decided not to produce silk, and the gypsy moths were released into the wild, where they reproduced and in the caterpillar stage proved to be very hungry insects. They will eat the leaves of over 500 species of trees!

© Frank Schaffer Publications, Inc.

Name_____

FISHY FACTS

If you do not know the answers to these questions, find them in a guidebook or encyclopedia. Some of the questions are thought-provoking questions, so think hard!

1. THE STARFISH AND REGENERATION: Asterid seastars, commonly known as starfish, feed on oysters. Several years ago, oyster fishermen tried to lessen the seastar population by cutting each star into small pieces. Unfortunately, seastars can regenerate or grow a whole new seastar from just one small piece, even just one arm. What do you suppose happened when the fishermen chopped up hundreds of starfish?_____
_____

2. FLAT AS A FLOUNDER: Flounders start life looking like normal baby fish, with one eye on each side of their heads. As they grow, one eye migrates to the other side of the head. The flounder, which is now blind on one side, tilts its body so that it can see. Gradually, its body flattens, the side with the seeing eye becomes colored like its back, and it sinks to the bottom of the sea looking like a typical adult flounder. The other eye gradually migrates around to the top of the flounder's body. Look at a picture of a flounder against the ocean floor. What advantage is the flounder's shape and coloration?_____
_____

3. SPONGE CRAB: The sponge crab carries a piece of living sponge on its back. The sponge is a captive, but it does get carried to fresh feeding grounds as the crab travels around the sea floor. What advantage might the sponge be to the crab? _____
_____

4. CLEANER FISH: The cleaner wrasse cleans the fungus, parasites and disease off other fish. It has a special shape and coloration and makes a special movement to tell its customers that it is a cleaner fish. The saber-toothed blenny is a nippy look-alike. The blenny eats other fish. Considering this, how do the blenny's color and shape serve its needs? _____
_____

© Frank Schaffer Publications, Inc.   44   FS-10112 Hands-On Science Experiments

CORAL REEF WALL

Why do we need to protect the coral reefs?

WHAT YOU NEED: One large, deep rectangular or square pan, sand, water, a board cut one inch shorter than the width of the pan and large driveway gravel.

WHAT TO DO:
1. Pour water in the pan to a depth of two inches. Make an island of sand at one end of the pan. The sand must mound up above the water line.

2. Make a border of gravel around the island. The gravel should mound up to form a wall that stays just under the surface of the water.

3. Using the board, gently make waves in the water. What happens? _____
_____

4. Increase the size of the waves, being careful not to splash water out of the pan. What happens? _____
_____

5. Remove the gravel. Now, make gentle waves. What happens? _____
_____

6. Make bigger waves. What happens? _____
_____

THINK: What do coral reefs do for islands and beaches? _____
_____
_____

THE CORAL REEF

Pictured below are some of the organisms that interact with the coral reef and help to keep the balance in this delicate ecosystem. The numbers next to each organism correspond to a number next to a blank at the bottom of the page. Fill in each blank with the name of the appropriate organism.

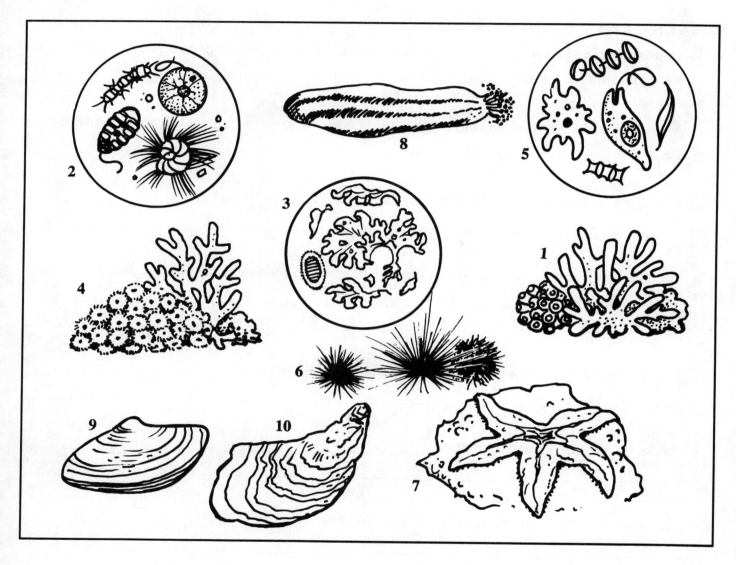

1. _____   5. _____   8. _____
2. _____   6. _____   9. _____
3. _____   7. _____   10. _____
4. _____

Name_____

TYPES OF MAMMALS I

Look at the pictures below. Classify the animals in the list in which they belong.

| MONOTREMES | MARSUPIALS | RODENTS |
Lay eggs	Have pouches	Gnaw
_____	_____	_____
_____	_____	_____
_____	_____	_____
_____	_____	_____
_____	_____	_____

Can you name other examples of these groups of mammals?

Name_____

TYPES OF MAMMALS II

Look at the pictures below. Classify these mammals under the title of the group in which they belong.

| CARNIVORES | AQUATIC MAMMALS | PRIMATES |
| Meat eaters | Live in water | Walk erect |

_____ _____ _____
_____ _____ _____
_____ _____ _____
_____ _____ _____
_____ _____ _____

Can you name more examples of these categories to fill in the extra blanks?

Name_____

THE SKELETON

Write the name of each bone in the appropriate blank on the left.

1. _____
2. _____
3. _____
4. _____
5. _____
6. _____
7. _____
8. _____
9. _____
10. _____
11. _____
12. _____
13. _____
14. _____
15. _____
16. _____
17. _____
18. _____
19. _____
20. _____
21. _____
22. _____
23. _____
24. _____
25. _____

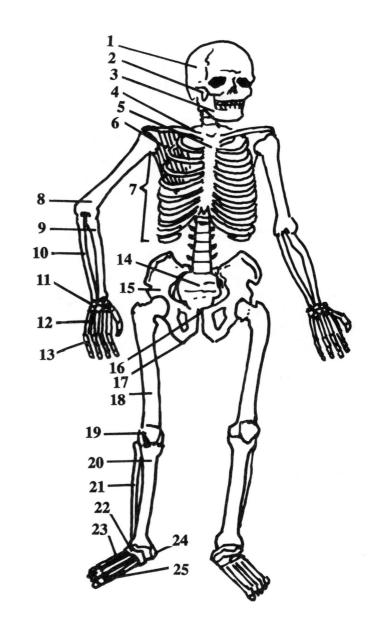

THINK: What is the purpose of the skeleton? _____
_____

© Frank Schaffer Publications, Inc.   49   FS-10112 HANDS-ON SCIENCE EXPERIMENTS

Name_____

FORENSIC SPECIALIST

Congratulations! You are the nation's leading forensic specialist. The police departments of Honolulu, Los Angeles, New York City and Chicago have asked you to assist them in solving four cases by identifying the remains they have found. Using your knowledge of the human skeleton, help them solve these four homicide and missing person cases. Write your solution in the blank below each case. Be prepared to defend your choice based on scientific evidence.

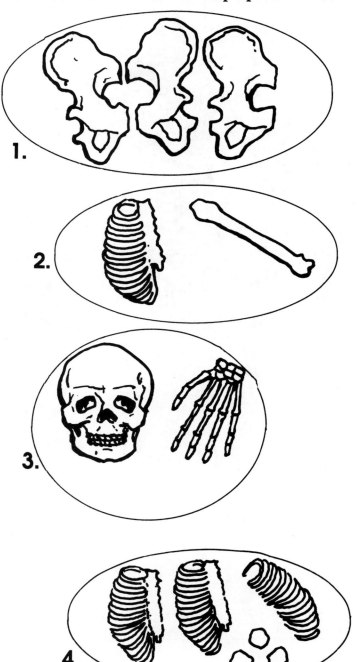

LOS ANGELES
These bones were found in a trash dumpster behind the restaurant, Le Posh. The police want to know if these are more likely to be the bones of the head waiter, Jacques Smith, who was feuding with the chef, or Mr. and Mrs. Ponsonby Smythe, the wealthy English couple who were last seen dining at table four.

_____

CHICAGO
The police have unearthed the bones pictured at the left. They think these are the key to the mysterious disappearance and suspected death of the drug runner, Notso Bryte. Could they be right?

_____

HONOLULU
The police found these bones on the beach. They don't know if they are the bones of the surfer Hy Wave or three men from the fishing tub Sunny Weather which has never returned to port. Which is more likely?

_____

NEW YORK CITY
The police found these bones buried in the concrete foundation of a demolished apartment building. Are they more likely to be the bones of the crime boss, Coss Anostra, or the cat burglars, Sy and Amy Fuzzifer?

_____

© Frank Schaffer Publications, Inc.   50   FS-10112 Hands-On Science Experiments

Name_____

DINOSAUR CELEBRITIES

Choose names from the Name Bank in the lower right hand corner to Identify the dinosaurs pictured below. Write the correct name in the blank with the number that corresponds to the number beside that dinosaur.

1. _____ 4. _____ 7. _____

2. _____ 5. _____ 8. _____

3. _____ 6. _____

Which dinosaurs were carnivores (meat eaters)? _____
_____
_____

Which dinosaurs were herbivores (plant eaters)? _____
_____
_____

NAME BANK

Allosaurus
Apatosaurus
Iguanodon
Stegosaurus
Tyrannosaurus Rex
Ankylosaurus
Anatosaurus
Triceratops

© Frank Schaffer Publications, Inc. FS-10112 Hands-On Science Experiments

Name_____

MAKING FOSSILS

Complete the experiment below to examine two different fossil formations.

WHAT YOU NEED: Plaster of Paris, two foil pie pans, modeling clay, water and a key.

WHAT TO DO:
1. Fill the bottom of one pie pan with modeling clay to a depth of one inch.

2. Press the key into the clay and remove it to leave an impression of the key.

3. Mix two cups of plaster of paris with water according to the package directions.

4. Pour a little of the plaster of paris mixture into the impression of the key and let it harden.

5. Pour the rest of the plaster of paris mixture into the other pie pan. Press your hand into the plaster of paris to make an impression and then remove it. Let the plaster of paris harden.

6. Carefully remove the hardened plaster of clay. This is an example of one way a fossil is formed. The plaster of paris cast of your hand is another.

THINK: Dinosaur bones and other fossils are layers of soil and sediment hardened by time. Why are some fossils found deeper down in the earth than others? _____
_____
_____
_____

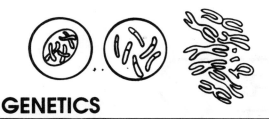

GENETICS

The way a person looks depends upon the genetic code he or she inherited from his or her parents. The children in a family do not necessarily look exactly like their parents. Look at the the characteristics of the parents in the two families below and see if you can tell what the children look like.

Dominant characteristics
curly hair–CC
dark hair–DD
brown, green or hazel eyes–BB

Recessive characteristics
straight hair–cc
red or light hair–dd
blue or grey eyes–bb

Mr. Smith has dark, straight hair and blue eyes. He is DDccbb. Mrs. Smith has blond, curly hair and green eyes. She is ddCCBB.

1. Sammy Smith is DdCcBb and so is Susie Smith. What color hair should they have? _____

2. Do you think their hair is straight or curly? _____

3. What colors might their eyes be? _____

Mr. Jones has dark, straight hair and blue eyes also. He is Ddccbb. Mrs. Jones also has blond, curly hair and green eyes. She is ddccBb.

4. Jack Jones is ddCcbb. What color is his hair? _____
    Is it straight or curly? _____
    What color might his eyes be? _____

5. John Jones is DdCcBb. What color is his hair? _____
    Is it straight or curly? _____
    What color might his eyes be? _____

6. Jill Jones is ddccBb. What color is her hair? _____
    Is it straight or curly? _____
    What color might her eyes be? _____

7. Jane Jones is Ddccbb. What color is her hair? _____
    Is it straight or curly? _____
    What color might her eyes be? _____

© Frank Schaffer Publications, Inc.   FS-10112 Hands-On Science Experiments

CHAPTER THREE

THE PLANET EARTH–GEOLOGY

Teacher Notes 55
Composition of the Earth–Rocks and Minerals 55
    Rocks and Minerals–How to Classify Them 62
    The Streak Test 63
    MOHS Hardness Test 64
    The Specific Gravity Test 65
Volcanoes and Earthquakes 57
    Volcano 66
    Epicenter of an Earthquake 67
Mountains 59
    Streams at Work 68
Groundwater 60
    Stalagmites and Stalactites 69
    Soil Permeability 70
Weather 61
    Dew Point 71
    Anemometer 72
    Weathervane 73

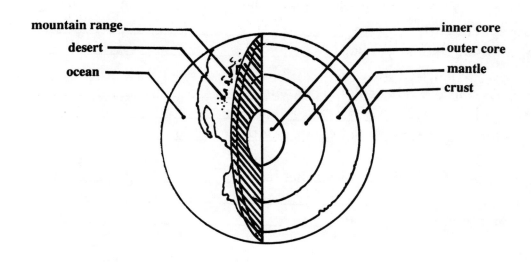

© Frank Schaffer Publications, Inc.

TEACHER NOTES

COMPOSITION OF THE EARTH–ROCKS AND MINERALS

Perhaps the best activity to acquaint students with the rocks and minerals of our planet is to create a rock collection. Rocks are readily available, easy to collect and fun to identify. Have students look for as many different kinds of rocks as possible. Then have them record and identify the location where they found each rock. Classify the rocks as sedimentary (fragmental or chemical), igneous (extrusive or intrusive) or metamorphic (foliated or nonfoliated). The student activity sheet ROCKS AND MINERALS–HOW TO CLASSIFY THEM (p. 62) is a chart for the students to use to classify these rocks. A good book about rocks and minerals, preferably one which has color pictures, would be a helpful addition to this activity.

The second, third and fourth student activity pages deal with three ways of identifying minerals. The following other tests should be explained and at least demonstrated to students: color, luster, crystal shape, cleavage, the flame test, the acid test, glow, magnetism, electrical conductivity, radioactivity and refraction. Discuss these tests and the reliability of relying on just one or two tests. If possible, have samples for students to test themselves. The activity sheets can be used with student-gathered samples, but depending on the region, some types of rocks may not be available to students.

The acid test and flame test may work well as demonstrations. For the acid test, have samples of calcite and galena as well as at least two other rocks/minerals. With an eye-dropper, place one drop of acid on the sample. Vinegar (acetic acid) may be used as the acid for either a demonstration or as a harmless acid for an individual student activity. Watch for a "fizz" reaction. For the flame test, the sample can be ground into a powder and placed directly in the flame of a Bunsen burner. Use a long platinum wire with a small loop at the flame end. Place the ground sample on the loop and insert into the flame. The test for electrical conductivity could also be performed as a demonstration.

The test for magnetism is simple and can be performed by the students. Have them touch a magnet to each sample and note any that are attracted to the magnet. The tests for luster, color, cleavage and crystal shape depend on observation and should be performed by the students as a regular part of identifying rock and mineral samples.

Activity Page 63 Teacher Key

THE STREAK TEST

Note: Try to have at least one sample for the students to test which will leave a streak that is a different color from the rock sample. Iron pyrite (fool's gold) is one good choice. A piece of porcelain bathroom tile will work quite well for a streak plate; simply use the unglazed (back) side.

Questions 1 through 10, the streak colors, should be answered based on each student's observations. The answer to the question at the bottom of the page should be "no" if you have provided the students with at least one sample which has a streak of a color different from the surface color of the rock/mineral.

Activity Page 64 Teacher Key

MOHS HARDNESS TEST

The full Mohs hardness scale is as follows:

Mineral	Hardness
Talc	1
Gypsum	2
Calcite	3
Flourite	4
Apatite	5
Feldspar	6
Quartz	7
Topaz	8
Corundum	9
Diamond	10

Provide students with 10 different mineral samples of varying hardness. The answers in the 10 blanks should correspond with the hardness of the 10 specimens as labeled.

THINK: The minerals might have a similar appearance, but the diamond is much harder.

Activity Page 65 Teacher Key

THE SPECIFIC GRAVITY TEST

Provide students with 10 different mineral samples. The answers for the 10 samples should be in accordance with each student's observations and consistent with the specific gravity of any known specimen.

© Frank Schaffer Publications, Inc.

VOLCANOES AND EARTHQUAKES

There are a number of ways to make model volcanoes. The method used in the first student activity has the advantage of being harmless to the student, the environment and is easy to clean when finished. In addition to this activity, discuss the structure of volcanoes and the basic structure types. (The three-category classification of shield volcano, cinder cone and composite volcano is considered outdated by geologists but is still used by some textbook editors. Those categories have been expanded to include many additional categories. The following list of several of the categories includes a definition and an example. Recent eruptions are noted. No note is made if the volcano is dormant or extinct.)

VOLCANO TYPES	
Tuff Ring:	is a wide, shallow crater. Buzzard Creek, Central Alaska
Maar:	is a wide, shallow crater (Tuff Ring) filled with water. Ukinrek, Alaska peninsula March 30 - April 9, 1977
Cinder Cone:	is a volcanic cone built of loose chunks of fragmental rock. North Pinhead (High Cascades), Oregon
Dome:	is a rounded, mushroom shaped protrusion of lava. Dome Complex–Dana, Alaska peninsula No known eruptions Dome Cluster–Augustine, Cook Inlet, Alaska 1812, 1883, 1935, 1963-64, 1976, 1986
Stratovolcano (Composite Cone):	is composed of both lava flows and fragmental material. Stratovolcano–Kiska, Western Aleutian Islands 1962, 1964, 1969. Stratovolcano Cluster–Tanaga, Central Aleutian Islands 1829, 1914
Caldera:	is a very large volcanic depression. Seguam, Central Aleutian Islands 1786-90, 1827, 1891, 1892, 1902
Shield Volcano:	is a broad, gentle slope built up from thin lava flows. Westdahl, Eastern Aleutian Islands 1964-65, 1978

Have the students draw or make models of different types of volcanoes or collect pictures of volcanoes and sort them into the different categories for a large poster or bulletin board. Volcanoes can also be classified by activity: active, dormant or extinct. Have the students find examples of each of these types and note the geographic closeness or distance from the students' homes or schools. The list above gives some idea of where to locate volcanoes in North America, but for more information, obtain a copy of *Volcanoes of North America, United States, and Canada* compiled and edited by Charles A. Wood and Jurgen Kienle.

Activity Page 66 Teacher Key — VOLCANO

WHAT TO DO:
3. The volcano erupts. The soda/vinegar fizzes out of the "volcano."

RESEARCH: Students should discover that the immediate area was affected by the fire, ash, etc.; but the effect was also felt globally because of the ash and gases in the atmosphere.

The January 1991 issue of *National Geographic Magazine* covered the eruption of Mount St. Helens in depth. This article would be an excellent resource for the students to use for further research.

The student activity on page 67 involves finding the epicenter of an earthquake. In addition to this activity, discuss the structure of the earth, including the following terms: crust, Moho, mantle, outer core and inner core. The following diagram could be reproduced for the students:

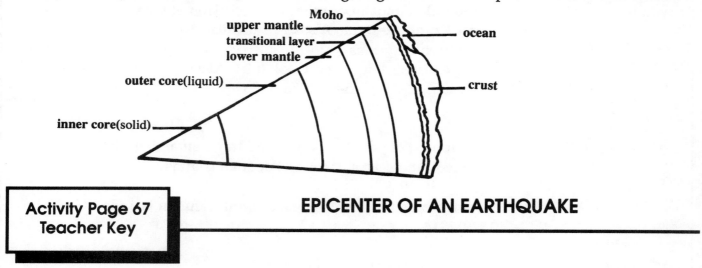

Activity Page 67 Teacher Key — EPICENTER OF AN EARTHQUAKE

1. Approximately 2 1/4 inches
3. Oakland, California, is the epicenter.

For a follow-up activity, have students research this earthquake. *National Geographic Magazine*, many other national magazines and all the major newspapers covered this earthquake. Discuss the likelihood of an earthquake in your area and what individuals should do in the event of an earthquake.

MOUNTAINS

Discuss the definition of the term *mountain*. Does a rise have to be a certain height to be a mountain? (No.) The Wachtung Mountains outside of New York City are less than 500 feet high. In order to better understand mountains, students may make a relief map of the local area. A relief map depicts land configurations with contour lines, shading or colors. The mountains can be made out of modeling clay or papier-mache and mounted on a base of foamcore or masonite. Papier-mache can be made by ripping newspaper into narrow strips and dipping the strips in liquid starch. Layer the strips to build up "mountains." Cover the final layer with smooth strips of starched paper and allow the ends of the paper to overlap the base.

Discuss the different kinds of mountains: depositional, volcanic, sand dunes, moraines (glacially deposited ridges composed of rock fragments), drumlins (long, narrow or oval smoothly rounded hills deposited by glaciers) and erosional, including mesas (flat topped rises totally or partially bounded by steep rock walls) and buttes (an isolated hill or mountain rising above the surrounding plain). Have students locate pictures of each for a bulletin board or poster of mountains.

Demonstrate the ease of forming sand dunes by having a student blow through a straw at a mound of sand in a large pan. The sand dunes in deserts and at beaches shift just as easily. Have a student pour water over the mound of sand; then discuss the purpose of sea oats and other natural grasses and plants on beach sand dunes.

To demonstrate the making of and to reinforce the use of topographical maps, make an irregular mountain of modeling clay in the bottom of a nine-inch by thirteen-inch (or larger) plastic dishpan or storage container. Use a sheet of plexiglass slightly larger than the container for the cover. Place the cover on the dishpan and have a student use a grease pencil to outline the base of the mountain on the plexiglass. Add one teaspoon of green or blue food coloring to one gallon of water. Pour the tinted water into the pan to a depth of one-half inch. Have a student draw around the base of the mountain again. Add more water to a depth of one inch and repeat the process. Continue to add water in one-half inch increments and draw around the perimeter of the mountain until the summit is reached. Lift the plexiglass off and show the final map of the mountain.

To demonstrate the geological effects of ice, have a student fill a small plastic container to the top with water and then seal it. Place it in the freezer overnight. Observe it the next day and discuss the results. Relate this to cracks in rocks and frost heaving.

STREAMS AT WORK (p. 68) demonstrates the work of streams: deposition and erosion. The water erodes the topsoil and deposits it at the mouth of the stream thus thereby forming a delta.

Activity Page 68 Teacher Key

STREAMS AT WORK

The answers should reflect the student's actual observation.
1. The soil should form an island or delta.
2. Some of the small gravel should wash away with the soil.
3. The large gravel should remain intact.

GROUNDWATER

Activity Page 69 Teacher Key — **STALAGMITES AND STALACTITES**

STALAGMITES AND STALACTITES is a simple activity in which the growth of crystals demonstrates the formation of stalagmites and stalactites in limestone caves. If the students have difficulty differentiating between stalagmites and stalactites, point out that *stalagmite* has a "g" for ground, the place from where it grows, and *stalactite* has a "c" for ceiling, the place from where it grows.

The answers should reflect the student's observations. The salt should form a small "stalactite" at the end of the string and a small "stalagmite" on the pan beneath the string. Placing the apparatus in a warm, sunny spot will hasten the process.

Activity Page 70 Teacher Key — **SOIL PERMEABILITY**

The SOIL PERMEABILITY activity is essentially the test used to decide whether soil will "perk" for country homes which have their own sewage systems. Soil which does not allow a certain amount of water to pass through it is not suitable for homes with septic tanks. Conversely, highly permeable soil allows minerals and toxic wastes at dump sites to leach into the soil and to be carried greater distances than in impermeable containers and less permeable soil. Discuss these points with students and have them take these facts into consideration when they test samples of their soil.

In addition to these activities, students may find it interesting to test for mineral content in water. Take a sample of local water and boil it in a clean beaker until all the water has evaporated. Instruct students to look for a white residue in the beaker. This is the deposit from minerals that were in the water. "Hard" water will produce a fairly significant deposit, and some communities near the ocean may have some salt in the water.

1. The water should flow through the jar of sand after the sand is saturated.
2. The water should flow through the jar of gravel.
3. The water should not permeate the clay.
4. The results will depend on the type of soil tested.
5. Gravel and sand are most permeable.
6. Clay is least permeable.

WEATHER

A weather station is an excellent way to acquaint students with weather and its effects. Use a bulletin board or a large piece of poster board to record weather information, including temperature, barometric pressure, wind velocity and direction, cloud cover, dew point, and past weather (from old television, radio or newspaper weather reports). Keep a record of the weather over a period of several weeks.

Activity Page 71 Teacher Key — DEW POINT

WHAT YOU WILL NEED:
The easiest "nylon net" bag is the toe end of a pair of nylon stockings or pantyhose. Cut twelve inches of the hose, fill the toe with ice cubes and drape it over the side of the can. The wet stocking will tend to cling to the side of the can, automatically anchoring the ice cubes.
  5. The temperatures should reflect actual observation.
  6. The temperatures should reflect actual observation.
RESULTS:
  1. Yes. The dew point varies with the air temperature.
  2. Yes. The dew point varies with varying weather conditions.

Activity Page 72 Teacher Key — ANEMOMETER

The anemometer does not have a calibrated gauge to measure actual wind velocity. Students can record the wind speed as slow, moderate or fast and check weather reports for actual wind velocity. A wind sock can be also used to judge both velocity and direction of the wind.

Activity Page 73 Teacher Key — WEATHER VANE

The weather vane is made entirely from trash. Scrap lumber can be obtained from lumberyards for the base. Usually lumberyards are willing to give free scraps for student projects. The base can be any piece of flat board not larger than eight inches square. The length of the nail is determined by the thickness of the board and the height of the spool. The nail needs to extend above the board but not above the spool when the spool is mounted on the nail. Be certain that the glue chosen for this project will glue two pieces of plastic together. The pattern for this project is on page 74. Have students cut the pattern out and tape the whole pattern to the milk jug before cutting. Trace around the pattern with a marker to leave guidelines. When the weather vane is completed, it is portable and can be placed in any open area.

© Frank Schaffer Publications, Inc. FS-10112 Hands-On Science Experiments

ROCKS AND MINERALS–HOW TO CLASSIFY THEM

Sedimentary Rocks:

| FRAGMENTAL Fragments that have eroded from other rocks | CHEMICAL Formed from minerals dissolved in water |

1. Conglomerate–large fragments bonded by sand
2. Breccia–sharp fragments usually bonded by clay
3. Medium fragments–example: sandstone
4. Fine fragments–example: shale, mudstone

Examples: halite (rock salt), limestone

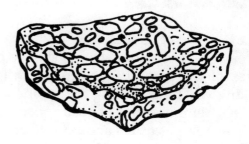

Igneous Rocks:

| INTRUSIVE Solidify above the earth's surface | EXTRUSIVE Solidify below the earth's surface |

Examples: granite, diorite, gabbro, peridotite, syenite

Examples: basalt, felsite, obsidian, pumice, scoria

Metamorphic Rocks:

| FOLIATED Mineral crystals are in layers; look striped or banded | NONFOLIATED Not banded or layered; break into sharp, angular pieces |

Examples: Slate–can be broken into sheets. Shist–can be divided into thin sheets. Mica schist can be separated with the fingernail. Gneiss–appears very banded.

Examples: marble, quartzite

THE STREAK TEST

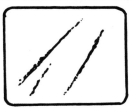

WHAT YOU NEED: A piece of unglazed porcelain and ten different rock samples. The porcelain is your streak plate.

WHAT TO DO:
1. Make a tag for each rock sample and give each a number. Attach the tag to the rock.

2. Rub sample #1 across your streak plate. At the bottom of the page are ten blanks. In the first blank, record the color of the streak which the rock leaves on the plate.

3. Repeat Step 2 with the rest of the rock samples. Record the color of the streak next to the correct number at the bottom of the page after you test the sample.

RESULTS:

1. _____
2. _____
3. _____
4. _____
5. _____
6. _____
7. _____
8. _____
9. _____
10. _____

Did all of the samples have a streak the same color as the rock itself? _____

Name_____

MOHS HARDNESS TEST

HARDNESS	EXAMPLE	PROCEDURE
1	Talc	Easy to scratch with fingernail
2	Gypsum	Hard to scratch with fingernail
3	Calcite	Can scratch with penny, not with fingernail
4	Fluorite	Can scratch with steel knife, not with penny
5	Apatite	Hard to scratch with steel knife
6	Feldspar	Can scratch glass, cannot be scratched with steel knife
7	Quartz	Can easily scratch a steel knife and glass

WHAT YOU NEED: A penny, a steel knife, a piece of glass, and minerals.

WHAT TO DO:
1. Make a tag for each mineral sample and give each a number. Attach the tag to the mineral.
2. Test mineral number one by first scratching it with your fingernail. If it scratches easily, write the number one for its hardness in the blank next to the number one below. If it is hard to scratch but you can scratch it, write two in the blank. If it will not scratch, try to scratch it with the penny. Keep on testing the mineral until it can be scratched. Always start by scratching with a fingernail and progress to scratching with a steel knife.
3. Test each of the other minerals in turn, and record their hardnesses in the blanks below.

NOTE: Some minerals fall in between two numbers. If the hardness is between 4 and 5, for example, label it 4 1/2.

SAMPLE	HARDNESS	SAMPLE	HARDNESS
1.	_____	6.	_____
2.	_____	7.	_____
3.	_____	8.	_____
4.	_____	9.	_____
5.	_____	10.	_____

THINK: Why would this test be useful in identifying two minerals such as quartz and diamond?

_____

_____

Name_____

THE SPECIFIC GRAVITY TEST

WHAT YOU NEED: Mineral specimens, a hanging scale, thread and a container of water.

WHAT TO DO:
1. Prepare ten labels or "placemats" for the minerals. Place these under or beside each specimen after you finish weighing it. Record the dry and wet weights next to the number which corresponds to the label of the specimen; then figure the specific gravity.

2. Tie one end of the thread around the specimen and the other to the scales. Weigh the specimen in the air. This is the dry weight.

3. Submerge the specimen in the container of water, but do not let it touch the container. Keep the scales out of the water. Weigh the submerged specimen. This is the wet weight.

4. To find the specific gravity, subtract the wet weight from the dry weight. Divide this amount into the dry weight. (dry weight - wet weight = weight loss; dry weight ÷ weight loss = specific gravity)

SPECIMEN	DRY WEIGHT	WET WEIGHT	SPECIFIC GRAVITY
1.	_____	_____	_____
2.	_____	_____	_____
3.	_____	_____	_____
4.	_____	_____	_____
5.	_____	_____	_____
6.	_____	_____	_____
7.	_____	_____	_____
8.	_____	_____	_____
9.	_____	_____	_____
10.	_____	_____	_____

Name_____

VOLCANO

WHAT YOU NEED: Pie pan, small jar, modeling clay, vinegar, one teaspoon baking soda and 1/2 teaspoon red food coloring.

WHAT TO DO:
1. Place the jar in the center of the pan. Carefully mold the modeling clay around the outside of the jar to form the outside of the volcano. Leave the mouth of the jar uncovered.

2   Fill the jar with vinegar to 1/8 inch of the top. Add the food coloring.

3. Add the baking soda. What happens? _____
_____

RESEARCH:
Read about Mount St. Helens. How much area was affected by the volcano? Were the people who lived near the volcano the only ones who were affected? _____
_____
_____
_____
_____

© Frank Schaffer Publications, Inc.   FS-10112 Hands-On Science Experiments

Name_____

EPICENTER OF AN EARTHQUAKE

On the map below, find the epicenter of an earthquake that is 650 miles from Seattle, Washington, 180 miles from Reno, Nevada, and 400 miles from Las Vegas, Nevada.

WHAT YOU NEED: A compass, a ruler and a pencil

WHAT TO DO:
1. If Seattle is 650 miles from the epicenter of the earthquake, how many inches is that according to the scale on the map below?_____ Fix your compass to that measurement. Using Seattle as the center, draw a circle around Seattle with that measurement as the radius.

2. Do the same for Reno and Las Vegas.

3. The place where the circles intersect is the epicenter. What is the name of that town?_____

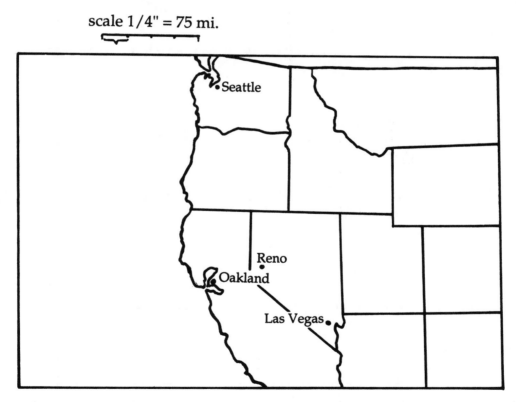

INTERESTING FACT: This is the site of an actual earthquake. It occurred on October 17, 1989, at 5:04 p.m. The epicenter of the actual Oakland or Loma Prieta earthquake was approximately 50 miles southeast of Oakland.

STREAMS AT WORK

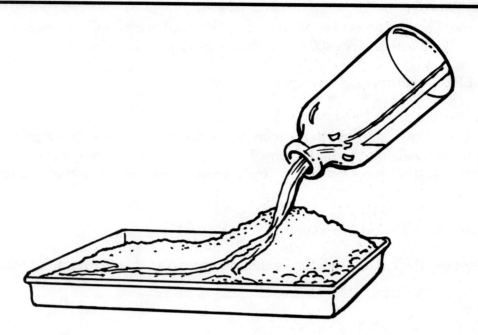

WHAT YOU NEED: A large pan, large driveway gravel, small aquarium gravel, topsoil and a large pitcher full of water.

WHAT TO DO:
1. In one end of the pan, put a layer of large gravel. Cover this large gravel completely with a layer of small gravel. Put the topsoil over the small gravel. Cover the small gravel completely. Pat the soil down.

2. Pour the container of water onto the topsoil. Pour it fast enough so that it erodes or washes away some of the topsoil and forms a small river.

RESULTS:
1. What happens to the soil that is washed away? _____

2. What happens to the small gravel? _____

3. What happens to the large gravel? _____

Name_____

STALAGMITES AND STALACTITES

WHAT YOU NEED: A pint plastic container with a leak-proof lid, 1/4 cup of salt, warm water, an eight-inch cotton string, a pin, three tall bottles of equal size and a pie pan.

WHAT TO DO:
1. Poke a hole in the lid of the container just large enough to thread the string through it. Knot the string on the inside of the lid.

2. Place the salt in the bottom of the container and add enough warm water to fill it. Stir the mixture until the salt is completely dissolved.

3. Place the lid on the container with the long end of the string extending from it. Turn the container upside down over the pie plate. Place the container on soft drink bottles or some other objects, making the container level and elevated above the pie plate so that the bottom of the string is two inches from the pie pan.

4. Check your experiment daily and record the results below.

RESULTS:

DAY 1 _____

DAY 2 _____

DAY 3 _____

DAY 4 _____

DAY 5 _____

Name_____

SOIL PERMEABILITY

WHAT YOU NEED: Four identical glass jars, sand, clay, gravel, soil from your area and water.

WHAT TO DO:
1. Fill the glass jars to one inch from the top, putting sand in one jar, clay in another, gravel in another and your soil in the fourth jar.

2. Label the jars with the type of soil in each jar.

3. Pour water into each jar filling it to the top. Observe and record your results below.

RESULTS:
1. What happens to the water in the jar containing sand? _____
_____

2. What happens to the water in the jar containing gravel? _____
_____

3. What happens to the water in the jar containing clay? _____
_____

4. What happens to the water in the jar containing your soil? _____
_____

5. Which soil is the most permeable? _____

6. Which soil is the least permeable? _____

Name_____

DEW POINT

WHAT YOU NEED: An empty tin can, a thermometer, nylon net, ice cubes and lukewarm water

WHAT TO DO:
1. Fill the tin can 3/4 full of lukewarm water.

2. Place the thermometer against one side of the can.

3. Place the ice cubes in the nylon net and pull the net around the ice cubes to form a bag that contains the ice cubes.

4. Place the bag of ice cubes in the can on the side opposite the thermometer. Do not let the ice cube bag touch the thermometer.

5. Watch for "dew" or moisture to form on the outside of can. When it does, check the temperature of the water. What is it? _____
   This is the dew point. What is the air temperature? _____

6. Repeat this experiment for ten days. Record the temperatures below.

	Air Temperature	Dew Point		Air Temperature	Dew Point
Day 1	_____	_____	Day 6	_____	_____
Day 2	_____	_____	Day 7	_____	_____
Day 3	_____	_____	Day 8	_____	_____
Day 4	_____	_____	Day 9	_____	_____
Day 5	_____	_____	Day 10	_____	_____

RESULTS:
1. Is the dew point different at different air temperatures? _____

2. Is the dew point different when the weather is different? _____

ANEMOMETER

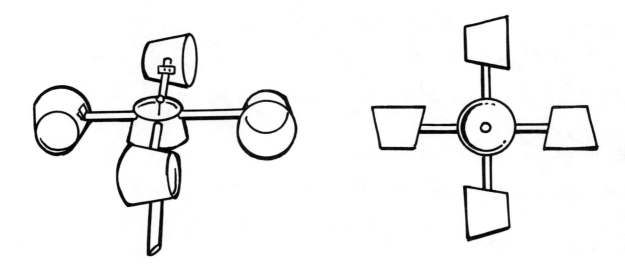

WHAT YOU NEED: One long florist's straight pin, one dowel, two drinking straws, five paper cups, masking tape and scissors.

WHAT TO DO:
1. Cut the top half off each cup. Cut a small hole in the side of four of the cups, just large enough so that you can fit the end of a drinking straw through it.

2. Cut four holes in the side of the fifth cup. The holes should be opposite each other, so the drinking straws can be threaded through the holes at right angles to each other. Thread the straws through the holes before you proceed to the next step.

3. Place one cup on each end of both drinking straws. Make sure that the cups are facing the opposite way from each other. One cup should be upside down and one right side up as they lie on a table. Fasten the cups onto the straws with masking tape by folding a small piece of the straw against the side of the cup and taping it.

4. Stick the long florist's pin through the paper cup and the two drinking straws and anchor it in the dowel. The cups on the ends of the straws should be able to move easily, like a sideways pinwheel.

5. Take your anemometer outdoors and poke the end of the dowel in the ground in an open space where the wind can reach it from all sides.

Name_____

WEATHERVANE

WHAT YOU NEED: A one gallon plastic jug, an empty thread spool, a piece of board, a long nail, a hammer, tape, a marker, scissors, glue and acrylic paint.

WHAT TO DO:
1. Cut out the patterns on the next page. Tape the patterns onto the jug as shown in the diagram. Trace around the pattern pieces with a marker before cutting the pieces from the jug. Pay special attention to the "cut here" sections. Do not make the cuts any longer or any shorter than the marks.

2. Nail the long nail through the center of your board. Place the empty spool on top of the nail. The nail should not extend beyond the top of the spool.

3. Assemble the weather vane according to the diagram below. Glue the circle to the spool and allow the glue to dry before starting step 4.

4. Paint the weathervane. Paint the points of the compass on the board which you are using as a base.

5. Mount the weathervane on its base by folding on the dotted lines and then taping the tabs onto the circular base. Take it to an open area where it will get the full effect of the wind. Use a compass to determine the proper direction for the weathervane. The N should point due north.

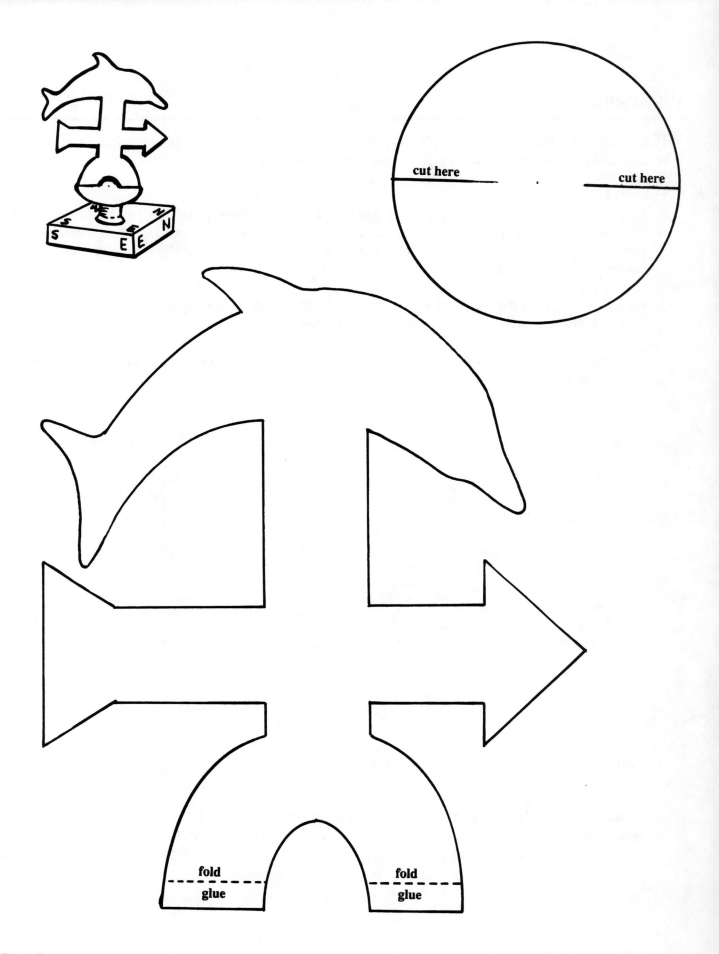

CHAPTER FOUR

BIOLOGY AND THE ENVIRONMENT

Teacher Notes	76
Water	76
Water Uses	83
Take the Salt Out	84
Cleaning Water	85
Balance of Nature	77
Food Webs	86
Interdependence	87
Endangered Species	88
Extinct Species	89
Conservation	79
Tree Conservation	90
Soil Conservation	91
Alternative Energy Sources	92
Solar Energy	92
Wind Power	93
Pollution	80
Smog	94
Water Pollution	95
Recycling	81

© Frank Schaffer Publications, Inc.

TEACHER NOTES

WATER

Discuss the need for water, its uses and availability. Note that all living things need water, not just people. Discuss dehydration in people (*Reader's Digest*, July 1992, has an excellent article). The first activity sheet, WATER USES, addresses the need for clean water.

Activity Page 83 Teacher Key — WATER USES

1. Firemen could use water polluted by inflammable pollutants.
2. Farm crops need unpolluted water.
3. A hydroelectric plant could use polluted water if it did not clog the equipment.
4. We need unpolluted water to drink.
5. We need unpolluted water to wash cars to protect their paint.
6. We need unpolluted water to wash dishes.
7. We need unpolluted water for boating.
8. We need unpolluted water for swimming.
9. We need unpolluted water for fishing.

Activity Page 84 Teacher Key — TAKE THE SALT OUT

This activity demonstrates the solar method of desalinating water. Discuss the practicality of this method on a large scale in terms of the amount of space and time involved versus the expense of boiling water.

Note on Materials: If you have a plastic container with a lid that curves slightly to form a saucer or shallow bowl, omit the pie pan and place the smaller container on the lid with the bottom of the plastic container over it.

4. The water on the lid container should taste fresh. The water in the small container should taste very salty. Water should accumulate on the lid container and run down the sides. If the lid container is sloped enough, fresh water may accumulate on the bottom pie pan or lid.

Activity Page 85 Teacher Key — CLEANING WATER

Prior to doing this activity, discuss these types of pollutants: chemical, sewage, trash, nuclear, soil. Discuss methods of cleaning water. This would be a good introductory activity to use prior to visiting a sewage treatment or water filtration plant.

5. The water running through the driveway gravel should be free of large debris, but muddy.
6. The water running through the aquarium gravel should be less muddy and free of large debris.
7. The water running through the cheesecloth should be less muddy than the other containers and free of large and small debris.

© Frank Schaffer Publications, Inc.

BALANCE OF NATURE

The activity sheet, FOOD WEBS, consists of two food webs, one containing primarily land plants and animals and one containing primarily marine plants and animals. All of the plants and animals eventually become part of the soil and furnish food for the plants, but the arrows for this are omitted on these webs. Accept either way as correct.

Activity Page 86 Teacher Key — **FOOD WEBS**

Activity Page 87 Teacher Key — **INTERDEPENDENCE**

INTERDEPENDENCE is a series of thought-provoking questions and ideas on the interrelatedness and interdependence of living things. Either prior to using this page or as a follow-up activity, discuss the concept of interdependence and the ways in which living things depend on one another.

1–10. Accept any reasonable answer and explanation. Discuss the virtues of each student's choices.

Spiders eat other insect pests. The garden spiders are particularly useful. Bees pollinate flowers, crops and fruit trees. The honeybees are also useful for food (honey). Hawks keep the rodent population under control. Algae provide food for animals higher on the food chain. Earthworms aerate the ground and allow moisture to penetrate dry earth more easily.

© Frank Schaffer Publications, Inc.   FS-10112 Hands-On Science Experiments

Activity Page 88 Teacher Key

ENDANGERED SPECIES

ENDANGERED SPECIES addresses the value and importance of endangered species. Use this as an introductory or follow-up activity for a discussion of endangered species and the reasons why we protect them. In conjunction with this discussion, consider the increased deer population in the Northeast and the increased white mountain goat population in Olympic National Park (see "Revenge of the Sea Lions" in *Reader's Digest*, July 1992). What would be the effect of a very large population of each of the endangered species? EXTINCT SPECIES addresses the opposite side of the issue. What are we missing and what do we lose when a species becomes extinct?

1. The whooping crane might be important to us for beauty, for food or for scientific reasons. The whooping crane suffers from loss of habitat and hunters.
2. The bald eagle is primarily valuable for its beauty and place of honor as our national symbol, but it also has scientific value. Like the whooping crane, it suffers from loss of habitat and hunters.
3. The jaguar was most widespread in Mexico. It became endangered because it was hunted for its fur and its beauty as a trophy.
4. The alala, or Hawaiian crow, has become endangered because of disease brought by other animals which were deliberately transported to Hawaii. It is valuable particularly as a scientific curiosity. It is still revered by some natives, who value it for its uniqueness: it eats its food with its feet and sounds more like a tiger than a bird.

Activity Page 89 Teacher Key

EXTINCT SPECIES

1. The dodo was important as a food source and possibly for its scientific value.
2. The Carolina parakeet was valued for its feathers. It also would have been of scientific value for its behavior. The birds were easy to kill because they flocked back to a hurt bird.
3. The Tasmanian wolf was valuable for its beauty and possibly for science.
4. The passenger pigeon was valuable for food and possibly for science.

CONSERVATION

TREE CONSERVATION demonstrates two methods of harvesting trees: selective cutting and total deforestation. SOIL CONSERVATION demonstrates the effects of contour plowing, terracing and vertical row planting. Grass seeds are suggested for these experiments because they are fast growing and easy to obtain; but any small, fast growing seed will work.

| Activity Page 90 Teacher Key | **TREE CONSERVATION** |

5. The soil should erode and wash to the bottom of the pan.
6. The seedlings should hold the soil in place.
7. Totally deforested areas can be replanted with seedlings. (Explain that many times lumbermen do this when they cut all of the trees in one area.)

| Activity Page 91 Teacher Key | **SOIL CONSERVATION** |

Terracing and contour plowing both help to prevent erosion. Also discuss strip planting, the process of planting different crops in strips or sections. (This is useful when farmers are planting row crops like corn on sloping land. The corn must be carefully weeded and the ground tends to erode. By planting bushy, cover crops in alternating strips with the corn, the erosion is lessened.)

ALTERNATIVE ENERGY SOURCES

The activity, SOLAR ENERGY, explores ways we can use the sun for heating. Before students perform this activity, discuss conventional heating methods. How do we usually warm water? The air in our homes? What natural resources do we expend for heat?

| Activity Page 92 Teacher Key | **SOLAR ENERGY** |

3. The temperatures of the water in the two pans should be the same.
4. The temperature of the water in the sun should rise more rapidly and to a higher degree than the water in the shade. The answers should reflect student observation.
5. The sun-warmed brick should be hotter to touch.

THINK: Accept any reasonable answer. The sun's energy could be used to heat water, homes, greenhouses, etc.

Follow this activity sheet with a research project. Study the designs of houses and heating systems and determine which ones are energy efficient, which ones are not energy efficient and discuss ways to improve efficiency. Design a solar house with a solar heating system.

Activity Page 93
Teacher Key

WIND POWER

The activity, WIND POWER, applies the energy of the wind to move a small object. Discuss other practical applications for wind energy: windmills, wind generators, water and ice sailing vessels and gliders.

1. The force of the air (wind) is moving the car.
2. Accept any reasonable answer (electricity, windmills, sailing, hanggliding).

POLLUTION

SMOG and WATER POLLUTION demonstrate the pollution of air, soil and water. The first activity shows the formation of smog. The second shows the way toxic chemicals can leach into the soil and water supplies. Discuss the different kinds of chemicals and pollutants that are deposited in the soil and how they got there: animal and human sewage, chemicals from farm crops, deliberate dumping of household toxic chemicals (oil, paints, antifreeze), deliberate dumping of industrial waste, unintentional leaks of waste, etc. For additional research and discussion, use the following topics: How does mercury get in tuna fish? What are some commonly used toxic pesticides? What would be the best way to dispose of used car oil, old paints, paint thinners and pesticides? Can any toxic chemicals be recycled?

Activity Page 94
Teacher Key

SMOG

The answers should reflect the student's observations.
3. The smoke should stay in the can because cold air is heavy.
4. The smoke should rise in the air because warm air is light.

Activity Page 95
Teacher Key

WATER POLLUTION

3. The food coloring mixes with or spreads into the soil. The surrounding soil should turn the color of the dye.
4. The water should carry the dye down to the bottom of the pan. The water should become colored with the dye.

THINK: The chemicals leach into the soil and water supplies. The chemicals could be put in special containers or recycled.

RECYCLING

The merits of recycling some things are debatable and in some cases, the cost of recycling may be more than that of creating anew, both in economic terms and in terms of natural resources. (Lynn Scarlett's "Don't Buy These Environmental Myths" from May 1992 *Reader's Digest* is a good article to discuss with students.) If you can take your students to a recycling plant, have them observe the kinds and amounts of energy being used and the recycling plant's effect on the environment. What follows is not a demonstration of a recycling plant or an experiment to demonstrate the value of recycling. Rather, here are some creative uses for trash.

USE THAT TRASH

1. Large plastic jar lids cleaned and with all removable paper and small pieces removed make good baby toys. Boil the tops to sterilize them for teething babies.

2. Jars with metal lids make good penny banks. Remove the lid and punch a slot in the lid with a hammer and a nail. Screw the lid back on and decorate it with acrylic paint or stickers.

3. Clear wide-mouth jars make dish gardens. Plant small plants such as dwarf violets or slow growing cacti in them. Decorate the mouth of the jar with a pretty ribbon.

4. Tin cans make pencil holders or cookie or candy tins for gift giving. Decorate them with paint or glue-ons.

5. Newspapers make papier mache when they are dampened with liquid starch or a mixture of flour and water or wheat paste. Make critters, piñatas, toys or Christmas decorations.

6. Gum and candy wrappers, particularly pretty foil wrappers, cut into small pieces make a lovely mosaic collage or glue-on decorations for those small pencil holders and gift tins in suggestion number four.

7. Large plastic milk jugs, bleach bottles, detergent containers or cat litter containers make good bird feeders. Rinse them out thoroughly and then cut an opening on each side of the container, leaving at least 1 1/2 inches untouched at the bottom. Fill with birdseed. Tie a strong cord or fasten a wire around the neck of the jug for attaching to a tree limb.

8. Cut magazine pages into one-inch by three-inch equilateral triangles. Dip each triangle into liquid starch or Elmer's glue and roll it tightly around a flattened paper clip, starting at the base of the triangle. Remove the dry beads and string them to make necklaces and bracelets.

9. Flip tops and bottle tops make Christmas decorations. Link flip tops from cans to form chains to hang on the Christmas tree. With a hammer and nail, punch a hole in the center of each bottle top. Paint the bottle cap with acrylic paints. String the caps on a piece of pretty ribbon tying a knot between each cap. Use a long strand as a chain or string two or three caps on a short ribbon and loop one end of the ribbon to hang it on a tree branch.

10. Egg cartons make "critters." Cut the egg cartons into two long "caterpillars" or one, two or three humped "bugs" (or cars or dinosaurs). Decorate with stickers, glue-ons, paint and glitter. Use pipe cleaners for antennae. Encourage imagination.

Have students think of other creative ways to use trash. Check the local library for hundreds of other ways to use trash. Artists are making sculptures from metal parts, mobiles from everything imaginable and pictures from lint from the dryer. Encourage your students to be tomorrow's creative geniuses with trash.

Name_____

WATER USES

We use water for so many things. We use it for recreation and transportation, for energy, to extinguish fires, as a cleaner and solvent, to grow food and to drink. Look at the pictures below. In the spaces below the pictures, tell which ones could use polluted water and which could not.

1

2

3

4
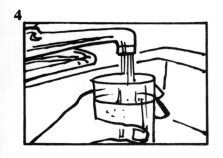

5

6

7

8

9

Name_____

TAKE THE SALT OUT

WHAT YOU NEED: A clear, quart size plastic or glass container, a pie pan, an eight-ounce shallow container, 1/4 cup of salt, and 3/4 cup of water.

WHAT TO DO:

1. Pour the water into the eight-ounce container. Add the salt and stir to dissolve the salt.

2. Place this small container on the pie pan and place them both in a sunny, warm window.

3. Cover the small container with the large container. The large container should completely surround the small container and sit inside the edges of the pie pan.

4. Observe the containers at the end of the day. Taste the water on the lid. What does it taste like? _____

   Taste the water in the container. What does it taste like? _____

   Is there water anywhere but on the lid or in the container? If there is, what does it taste like?

5. Pour off the water on the lid and measure it. How much water do you have? _____
   _____ Dry the lid off and replace it.

6. Repeat step 5 daily until the water in the small cup has evaporated and has left only salt. How long did this take? _____

Name_____

CLEANING WATER

WHAT YOU NEED: Three clean flower pots of the same size (the kind with one center drain hole and no saucer), driveway gravel, aquarium gravel, cheesecloth, water, soil which has some small roots, leaves, twigs, or other debris mixed in with it and a clean bucket or dishpan.

WHAT TO DO:

1. Label the glass jars: #1 = driveway gravel, #2 = aquarium gravel, and #3 = cheesecloth.

2. Fill the first pot with the driveway gravel. Fill the second pot with aquarium gravel. Line the third pot with three layers of cheesecloth. Make the cheesecloth come up over the lip of the pot as in the picture above.

3. Fill each of the glass jars 3/4 full of water.

4. Stir the soil to mix the debris evenly; then pour 1/4 cup of soil into each of the glass jars. Stir to mix it well.

5. Hold the flower pot with the driveway gravel over the bucket or dishpan. Pour the contents of jar #1 through the gravel and catch it in the bucket. Wash the jar and pour the water from the bucket into it. What color is the water? _____
How much debris is in it?_____
Wash the bucket or pan before starting Step 6.

6. Hold the flower pot with the aquarium gravel over the bucket or dishpan. Pour the contents of jar #2 through the gravel and catch it in the bucket. Wash the jar and pour the water from the bucket into it. What color is the water? _____
How much debris is in it?_____
Wash the bucket or pan before starting Step 7.

7. Hold the flower pot with the cheesecloth over the bucket or dishpan. Pour the contents of jar #3 through the gravel and catch it in the bucket. Wash the jar and pour the water from the bucket into it. What color is the water? _____

   How much debris is in it?_____

   Would you want to drink the water from either of these jars? _____

_____

© Frank Schaffer Publications, Inc.   FS-10112 Hands-On Science Experiments

Name_____

FOOD WEBS

The pictures below show the plants and animals in two food webs. The top picture shows land animals and plants, and the bottom one shows marine life and a human being. Draw arrows in each food web picture, from each creature to the plants and animals that depend on it for food. For example, in the top picture, the corn and deer are eaten by the person, so arrows should be drawn from the corn and the deer to the person.

© Frank Schaffer Publications, Inc.   86   FS-10112 Hands-On Science Experiments

Name_____

INTERDEPENDENCE

Animals and plants depend on one another for survival. If you were being sent to populate a new planet which had an abundant water supply, mineral resources, a good atmosphere and excellent climate conditions, what 10 plants and animals would you take with you? List them below and explain your reasons for taking each in the blank beside each number.

1. _____
2. _____
3. _____
4. _____
5. _____
6. _____
7. _____
8. _____
9. _____
10. _____

Many times animals and plants which are feared or considered pests are very necessary to our survival. Why are the following plants and animals useful?

Spiders _____

Bees _____

Hawks _____

Algae _____

Earthworms _____

Name_____

ENDANGERED SPECIES

Plants and animals are important to us for four basic reasons.

1. Beauty: We admire them. Example: roses.
2. Economic value: They are worth money to us; we make a living from them. Examples: silkworms, cotton.
3. Food value: We eat them or a part of them. Examples: rice, peanuts.
4. Scientific value: They are valuable to us now or may be valuable to us later for medicinal purposes, or they may be somehow necessary to our survival or well-being. Examples: manatees, dolphins, spiders.

Look at the animals pictured below. Next to each, write the reasons these animals or plants are important to us.

1. Whooping Crane

2. Bald Eagle

3. Jaguar

4. Alala

Name_____

EXTINCT SPECIES

The following animals are extinct. What did we lose when they disappeared? Look up these animals in an encyclopedia. Next to each animal, jot down reasons why you think it was important.

1. Dodo

_____
_____
_____

2. Carolina parakeet

_____
_____
_____

3. Tasmanian wolf

_____
_____
_____

4. Passenger pigeon

_____
_____
_____

Name_____

TREE CONSERVATION

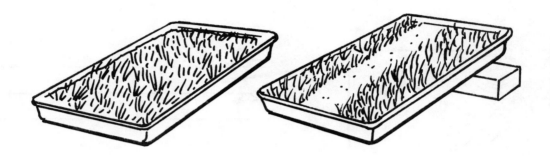

Two methods lumbermen use for harvesting trees are *selective cutting* and *total deforestation*. Try this experiment to see which method is better for the soil.

WHAT YOU NEED: Two 9" x 13" baking pans, potting soil, grass seeds and water.

WHAT TO DO:
1. Fill each pan with potting soil to a depth of one inch.

2. Sprinkle grass seeds evenly over the potting soil in each pan.

3. Water the pans and place them in a warm, sunny spot. Water the pans daily. When the seedlings are one inch high, you are ready to finish the experiment.

4. Pull all of the grass seedlings out of the center of one pan making a three-inch strip of bare soil from one end of the pan to the other. This is *total deforestation*.

5. Tilt the pan so the strip of bare soil runs downhill. Water the soil. What happens to the bare spot?_____

6. Pull only the tallest grass seedlings from the other pan. Pull seedlings from all parts of the pan, but be careful not to leave any bare spots. This is *selective cutting*. Water this pan. What happens? _____

7. What could be done to prevent erosion of land that has already been deforested? _____
   _____

Name_____

SOIL CONSERVATION

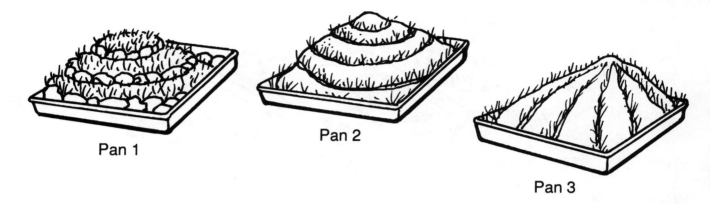

Pan 1

Pan 2

Pan 3

WHAT YOU NEED: Three large, flat pans; potting soil; grass seed; water; and small rocks or driveway gravel.

WHAT TO DO:
1. Label the first pan *Terracing*. Pour a 1/2-inch layer of potting soil in the center of the pan, about one inch smaller than the outside of the pan. Pack it down and place a wall of gravel or small rocks around it to keep it in place. Place a second 1/2-inch layer on top of the first, making it one inch smaller than the bottom layer. Place a wall of rocks around it. Make a top layer one inch smaller than the middle layer and fence it with a wall of rocks. Spread grass seed evenly over each layer.

2. Label the second pan *Contour*. Make a large mound of soil in the center of the pan. Firm the soil by patting it and carefully form ridges around it. Look at the picture for help. Make the ridges 1/2 inch apart. Plant grass seeds in the ridges.

3. In the third pan, make a mound of soil and pack it as in the second pan. In this pan, make ridges up and down the mound like the ridges on an umbrella. Look at the picture for help. Plant grass seeds in the ridges.

4. Water each pan to moisten the soil and place the three pans in a warm, sunny spot.

5. When the grass seedlings are one inch tall, water heavily.

RESULTS:
1. Which pan shows soil erosion? _____

2. What planting methods help to keep the soil from eroding? _____
_____

Name_____

SOLAR ENERGY

WHAT YOU NEED: Two bricks, two identical pans and water.

WHAT TO DO:
1. Place one brick in the sun and one in the shade. Make certain that the places chosen will remain sunny and shady for at least four hours each day and that the air temperature around the bricks is the same.

2. Pour an equal amount of cool water in each of the pans. Place one pan in the sun and one in the shade, the same as you did with the bricks. Label the sunny pan Number 1 and the shady pan Number 2.

3. Take the temperature of the water in each pan and record it below:

    Pan Number 1 _____   Pan Number 2 _____

4. Take the temperature of the water in each pan every hour for four hours. Record the results below:

Time	Pan Number 1	Pan Number 2
_____	_____	_____
_____	_____	_____
_____	_____	_____
_____	_____	_____

5. Touch each of the bricks after four hours. What has happened?_____
_____

THINK: How could the sun's energy be used for heating purposes?_____
_____

Name_____

WIND POWER

WHAT YOU NEED: Scissors, tape, a soda straw, a straight pin, a matchbox car and pliers.

WHAT TO DO:
1. Cut out the pattern at the bottom of this page. Cut all solid lines. Bend the pattern to make a pinwheel. Look at the pictures for help.

2. Cut the soda straw in half. Push the straight pin through the center of the pinwheel and then through the top of one half of the soda straw. Bend the end of the pin with the pliers to attach it to the straw.

3. Tape the bottom of the straw to a matchbox car so the pinwheel faces backward. Blow against the pinwheel.

RESULTS:
1. What is moving the car?_____

2. What are some other uses for wind power?_____
_____

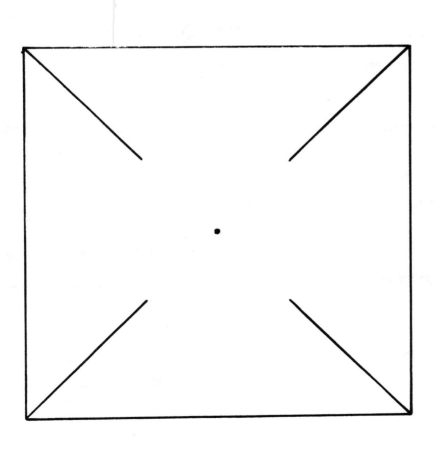
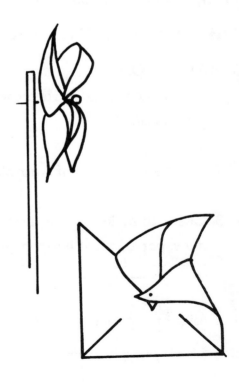

© Frank Schaffer Publications, Inc.   93   FS-10112 Hands-On Science Experiments

Name_____

SMOG

Smog occurs when pollutants are trapped on the ground by a warm upper air mass.

WHAT YOU NEED: Two empty tin cans, newspaper, one bowl of ice and matches.

WHAT TO DO:

1. Place one tin can in the bowl and surround it with ice. Make certain the inside remains dry. This is can #1.

2. Place a small wad of crumpled newspaper into the bottom of each can.

3. Check to make certain the air inside can #1 is cold. When the air is cold, light the crumpled newspaper. What happens to the smoke? _____
_____

4. Light the paper in can #2. What happens to the smoke? _____
_____

Name_____

WATER POLLUTION

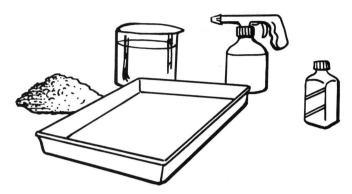

This experiment will show you how pollutants get into the soil and into the water supply.

WHAT YOU NEED: A large, flat pan, soil, blue or green food coloring, water and a mister or empty non-aerosol spray bottle.

WHAT TO DO:
1. Fill the pan with water to a depth of 1/2 inch.

2. Make a small island of soil in the center of the pan.

3. With your finger, make a small depression or hole in the center of the soil "island." Fill the hole with food coloring. What happens to the food coloring? _____
_____

   What happens to the soil? _____
_____

4. Fill the mister and spray the island. What happens? _____
_____

THINK:

What happens to toxic chemicals that are dumped on or in the ground? _____
_____

What could be done to prevent pollution of our soil and water from these chemicals? _____
_____

CHAPTER FIVE

ASTRONOMY AND SPACE

Teacher Notes 97

The Moon and the Solar System 97

    Viewing the Moon 101

    Moon Maps 102

    Making a Moon Globe 103

    Solar System Mobile 104

The Sun, The Stars and the Milky Way 98

    Homemade Planetarium 108

    Milky Way Spinner 109

    Binoculars Sun Viewer 110

Space 110

    Satellites in Orbit 111

    How to Feed an Astronaut 112

    Preparing for the Unknown 113

    An Experiment to Do in Space 114

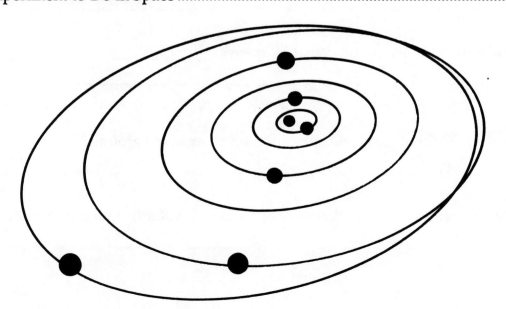

TEACHER NOTES

THE MOON AND THE SOLAR SYSTEM

VIEWING THE MOON

Two activities in this section suggest using a pair of binoculars for viewing astronomical bodies. A good pair of binoculars is almost as effective as a telescope (more effective than many small telescopes) because we see better when we can use both eyes even if the magnification is less. Use either binoculars or a telescope for these activities. Encourage students to become stargazers and remind them that many of the astronomical discoveries have been made by amateurs, not by professional astronomers with huge telescopes. For additional information, write to the Astronomical League. They will provide you (free of charge) with a copy of their newsletter, the address and contact number of the chapter nearest you, a booklet entitled *Understanding the Universe, a Career in Astronomy* and a sheet listing various astronomical resources.

Write to: The Astronomical League
            6235 Omie Circle
            Pensacola, FL 32504-7625

The first activity focuses on viewing our nearest neighbor, the moon. The activity suggests using binoculars, but naked eyesight or a telescope will work equally well. Note the times and moon phases suggested in this activity; some times and phases offer better viewing sights than others.

MOON MAPS

This activity involves a moon model, complete with craters, mountains, valleys and plains. The MOON MAPS on page 102 are to be used with both the first and the second activity. Run copies for students to use as guides. Pass them out after students have drawn their observations of the moon surface.

Activity Page 104 Teacher Key

SOLAR SYSTEM MOBILE

This mobile provides a good visual sense of the relative size of the planets. Each planet should be labeled with the name, the number of moons it has and any identifying information such as the number of rings, atmosphere, specific size, temperature, distance from the sun, etc. The mobile can then be laminated or covered with contact paper to ensure durability.

THINK: The model of the sun would be too large to fit on this mobile if it were in proportion to the size of the planets.

National Geographic Magazine and the *National Geographic Picture Atlas of Our Universe* by Roy A. Gallant offer excellent articles and pictures relating to individual planets and our solar system. *Our Universe* contains an interesting section on imaginary creatures for added student interest.

THE SUN, THE STARS AND THE MILKY WAY

The first activity in this section involves making simple star maps to shine on the ceiling like a planetarium. This is a good way to familiarize students with major constellations and stars. They should be able to recognize Ursa Major (Big Dipper), Ursa Minor (Little Dipper), Canis Major, Canis Minor and the zodiac constellations. In addition, they should be able to locate the North Star (Polaris). As they learn the major constellations, expand to the smaller ones. A study of Greek and Roman mythology, as suggested at the bottom of the activity on page 108, would be a good extension of this activity and would make it easier for the students to remember the constellations. Start with the stories about the following zodiac constellations:

Ursa Major, the Great Bear–the lover of Zeus who was changed into a bear by one of the gods
Ursa Minor, the Lesser Bear–son of Callistro
Canis Major, the Greater Dog–dog of Actaeon or Orion
Canis Minor, the Lesser Dog–favorite dog of Helen of Troy
Andromeda–daughter of Cassiopeia and Cepheus
Orion, the Great Hunter–loved by Apollo's sister, Artemis
Capricorn, the Sea Goat–half goat (head and body), half fish (tail)
Aquarius, The Water Carrier–represents the rainy season
Pisces, the Fish–Venus and Cupid in disguise to escape from the monster, Typhon
Aries, the Ram–Jason's prized golden fleece
Taurus, the Bull–Zeus's disguise to attract Europa
Gemini, the Twins–Sons of Leda.
Cancer, the Crab–Juno sent the crab to attack Hercules
Leo, the Lion–killed by Hercules as his first Labor
Virgo, the Virgin–Ceres, goddess of the harvest
Libra, the Scales–Astraea, goddess of justice
Scorpius, the Scorpion–sent to kill Orion
Sagittarius, the Archer–the centaur Chiron

Activity Page 109 Teacher Key   **MILKY WAY SPINNER**

The MILKY WAY SPINNER can be used as a model of our galaxy. Acquaint students with other galaxies and with major nebulae. The Andromeda galaxy and the Horsehead Nebulae in Orion are examples. Discuss the movement of galaxies and size of galaxies relative to the size of our planet.

SOMETHING ELSE TO DO: No. Galaxies are categorized as elliptical, spiral, barred spiral and irregular. Andromeda and the Milky Way are both spiral.

| Activity Page 110 Teacher Key | BINOCULARS SUN VIEWER |

This activity could just as easily be performed with a small telescope. Stress the importance of shielding the eyes from the direct rays of the sun. Viewing the sun through a telescope or binoculars is very dangerous, because the lenses concentrate the sun's rays and thus magnify the intensity and danger.

This activity is an excellent method of viewing sunspots. Have students observe the sun over a period of several days or every two to three days for a month and record the sun spot activity.

4. Sunspots seem to travel from west to east across the sun over a period of time. Other changes may be seen depending on sunspot activity. The answer should reflect actual student observation.

SPACE

The government and NASA provide information on space travel, NASA, space experiments and space information in specific fields of study such as biology and chemistry. Request information from:

Superintendent of Documents   and   National Aeronautics and Space Administration
U. S. Government Printing Office   Washington, D.C. 20546
Washington, D.C. 20402

The student activity, SATELLITES IN ORBIT, demonstrates the gravitational pull of planets and large masses in space. Discuss the value of satellites and the possibilities presented by satellites and space stations.

| Activity Page 111 Teacher Key | SATELLITES IN ORBIT |

3. The satellite is drawn to the earth. If this actually were a satellite, it would crash to Earth.
4. At a certain distance, the satellite is in orbit and can circle the earth.
5. The satellite gradually slows its orbit. Actual satellites can fall out of their orbits and be drawn back to Earth.

THINK:
1. Several problems confront scientists who try to place satellites in orbit: the force required to get beyond the earth's gravitational pull while keeping the satellite in orbital range, keeping a satellite in orbit, not interfering with other satellites which are already in orbit. . . . Accept any reasonable answer.
2. Accept any reasonable answer.
3. Accept any reasonable answer. Discuss the difficulties involved with using any place as a dump, including outer space.

© Frank Schaffer Publications, Inc.   FS-10112 Hands-On Science Experiments

HOW TO FEED AN ASTRONAUT

This activity does not have any discussion questions for the student to answer. In this activity, the focus is on the practicality of the student's idea. Does it work? Discuss the problems connected with living in space or on other planets. Challenge students to think of ways to meet the problems of space and then compare their answers with what NASA scientists and astronauts are already doing.

Activity Page 113
Teacher Key

PREPARING FOR THE UNKNOWN

PREPARING FOR THE UNKNOWN is an exercise in using the imagination. Discuss what is known about the other planets in our solar system and outer space and then ask for suggestions for other possibilities. Encourage creativity.
Accept any reasonable answers such as
1. Take clothing, plenty of provisions–food and water
2. gas masks, oxygen tanks
3. weapons, gifts

Activity Page 114
Teacher Key

AN EXPERIMENT TO DO IN SPACE

AN EXPERIMENT TO DO IN SPACE is meant to be an introductory activity. Student experiments are accepted by NASA for testing on space missions. Encourage students to develop original experiments which might be actually used by NASA. This experiment is just a prototype.

1. The spices should gradually sink; the oil should float on the top.
2. The oil, water and seasonings should separate.
3. The mixture should begin to separate again.
4. Accept any reasonable answer. The mixture might not separate.

© Frank Schaffer Publications, Inc.

Name_____

VIEWING THE MOON

WHAT YOU NEED: A pair of binoculars, paper and a pencil.

WHAT TO DO:
1. The best times to view the moon are in the latter waxing crescent through the first quarter phases as seen in the chart below. Go out on a clear night when the moon is in these phases and draw what you see on sheets of notebook paper. From what you record, draw a final map of the moon in the circle at the bottom of this page.

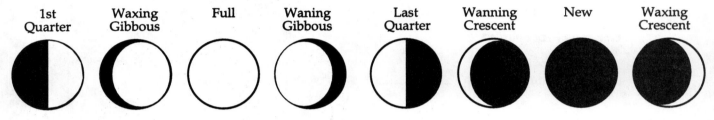

1st Quarter   Waxing Gibbous   Full   Waning Gibbous   Last Quarter   Wanning Crescent   New   Waxing Crescent

2. Label your Moon Map according to the map that your teacher gives you.

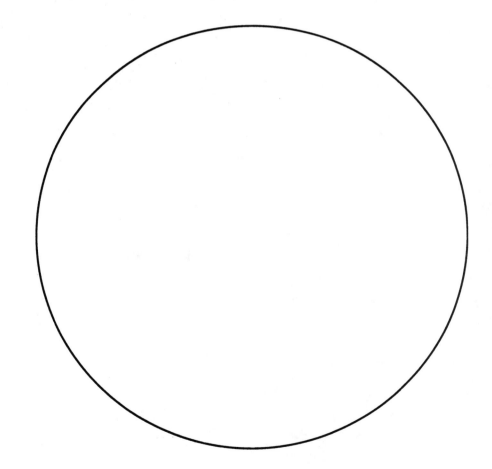

© Frank Schaffer Publications, Inc.

MOON MAPS

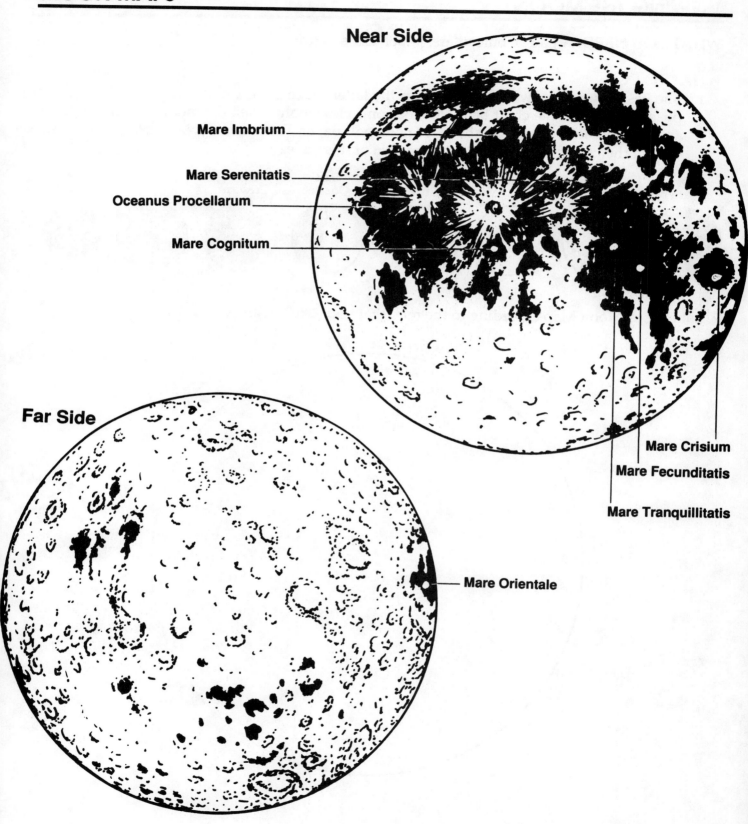

MAKING A MOON GLOBE

WHAT YOU NEED: One large round balloon, newspaper, liquid starch (or dry starch mixed with water to the consistency of Elmer's glue) and maps or a globe of the moon.

WHAT TO DO:
1. Inflate the balloon.

2. Tear the newspapers into long, narrow strips.

3. Dip the newspapers in the starch and lay them one at a time on top of the balloon. Each piece of newspaper should be placed perpendicular to the piece underneath it. The ends can overlap the underneath layers. Cover the whole balloon.

4. When the newspaper is six layers thick, start to build mountains and high places. To do this, make a wad of starched newspaper roughly the shape and size of the mountain, ridge or high place. Place it on the moon globe in the proper spot and attach it by laying flat strips of starched newspaper over the wad. Press the ends of the strips against the flat, low spots to hold the mountains in place.

5. After the globe is modeled, let it dry; then paint it with tempera or acrylic paint. Label the main landmarks, particularly the Sea of Tranquility where the *Apollo 11* astronauts Neil Armstrong and Edwin E. Aldrin, Jr. landed on July 20, 1969.

Name_____

SOLAR SYSTEM MOBILE

WHAT YOU NEED: Heavy thread, a needle, nylon thread or fishing line, wire (20 gauge wire will work), lightweight cardboard, glue, scissors, a single hole punch, ten index cards and paint, crayons or markers.

WHAT TO DO:
1. Color the planets.

2. Punch a hole in the center of the top of each index card so that you can hang the cards from the mobile.

3. On the first index card, write the word "SUN." On each of the other cards, write the name of a planet, the number of moons and any information that helps to describe the planet (size, distance from the sun, names of the moons, any known facts about temperature or atmosphere).

4. Cut the wire as follows:
    one piece seven inches long
    one piece twelve inches long
    one piece seventeen inches long
    one piece twenty-three inches long

5. Bend 3/8 inch of each end of each wire to form little hooks. Bend the center of each wire to form a loop.

6. Thread the needle with the heavy thread. With the needle, pierce the top of a planet (pierce it 1/2 inch from the edge if you have enough room or in the center if it is a tiny planet like Pluto). Pull the thread, being careful not to pull it all the way through the hole. Pierce the bottom of the index card and pull the thread part way through this hole. Tie the planet to its index card and cut the thread loose. Do this for all of the planets.

7. Cut a two-inch piece of wire and form a hook at each end. Cut twelve inches of nylon line and tie it to the center loop on the longest piece of wire. Suspend the wire and hook the shortest wire to the center loop. Hang the planet Pluto and its label from this hook. Hang Neptune and Uranus from each end hook of the longest wire.

8. Cut another foot of nylon line and tie one end of it to the center hook of the longest wire. Tie the other end to the center hook of the second longest wire. Cut more nylon thread and tie the next wire on the mobile, attaching it to the center loop of the wire above it.

9. Hang the planets and their cards on the end hooks. Hang the sun card from a thread or line tied to the center loop of the last wire.

THINK: Do you know why there is no model for the sun on this mobile? _____

_____

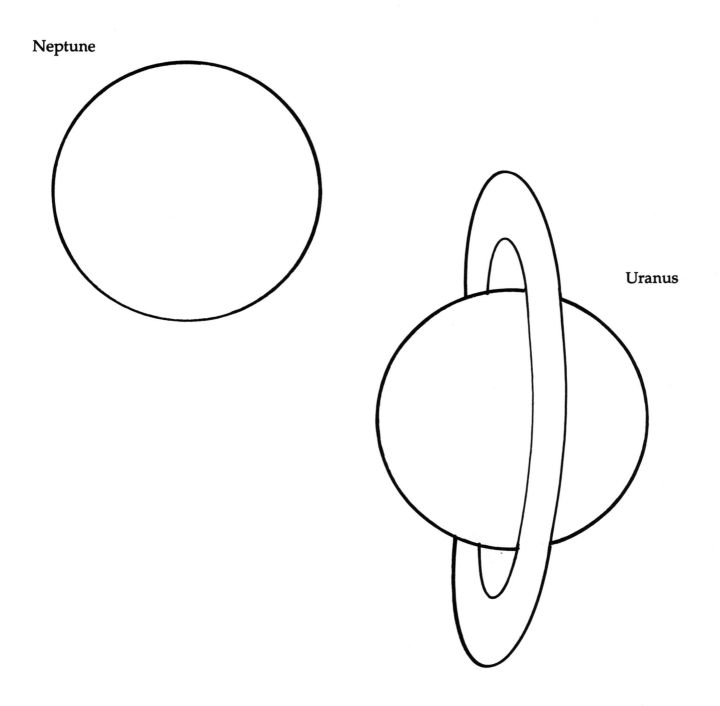

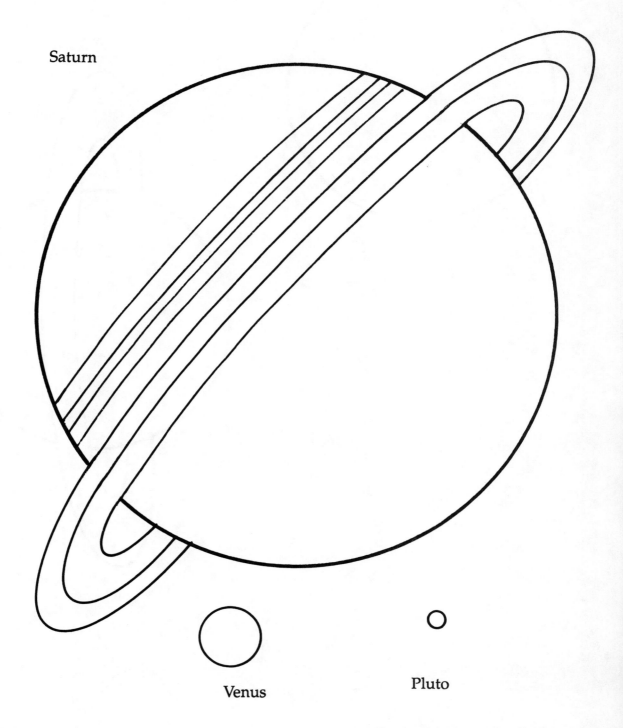

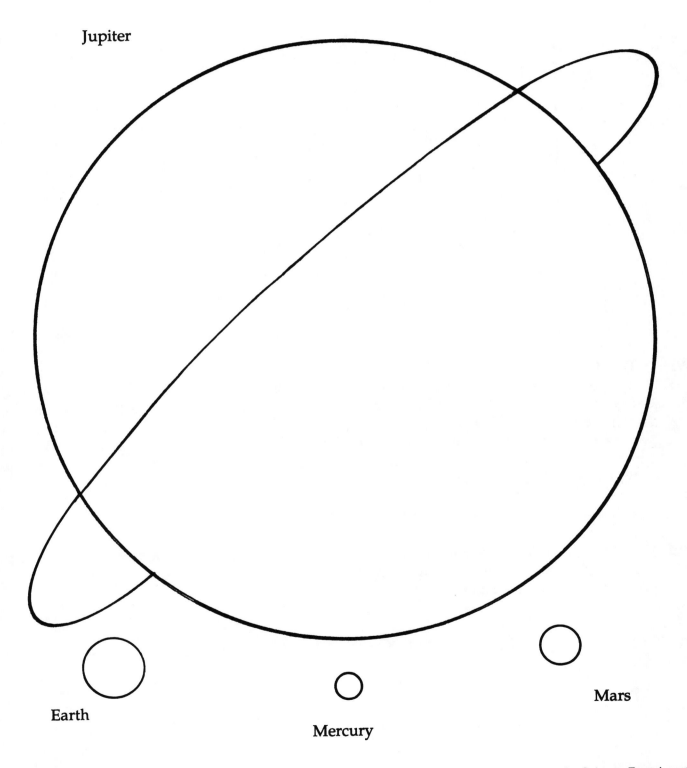

Name_____

HOMEMADE PLANETARIUM

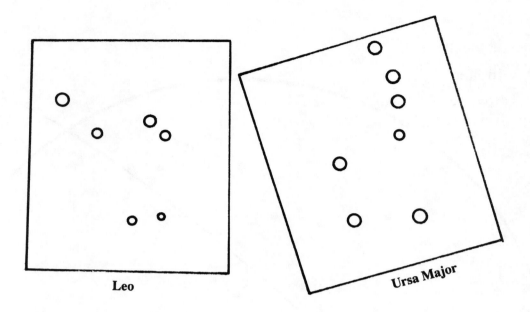

WHAT YOU NEED: A large nail, a pencil, sheets of construction paper and a flashlight.

WHAT TO DO:
1. With a pencil, make star diagrams of the major constellations: Ursa Major, Ursa Minor, Canis Major, Canis Minor, Andromeda, Orion, Capricorn, Aquarius, Pisces, Aries, Taurus, Gemini, Cancer, Leo, Virgo, Libra, Scorpio and Sagittarius. Place each constellation on a separate sheet of paper, marking a dot for each star in the constellation. Make the constellation as large as it can be and still fit on the paper. Label each constellation with the Latin name and its meaning.

2. Punch the nail through each dot to make a hole in the paper for each star.

3. When you have finished, darken the room. Hold the constellations above your head one at a time and shine the flashlight through the holes in the paper. The constellations will show on the ceiling just like real stars or the visual effects of a planetarium.

ANOTHER INTERESTING THING TO DO: Read stories of the constellations from Greek and Roman mythology.

Name_____

MILKY WAY SPINNER

WHAT YOU NEED: Scissors, a small nail, a large nail and a flashlight.

WHAT TO DO:
1. Cut out the disc at the bottom of this page.
2. Using the small nail, punch a hole through all of the dots except the center hole.
3. Push the large nail part way through the center hole. Tape the disc to the large nail if it slides too much.
4. Darken the room. Using the nail as a handle, spin the disc slowly above your head. Shine the flashlight through the holes. What you see on the ceiling represents our galaxy, the Milky Way.

SOMETHING ELSE TO DO: Make spinners and maps of different galaxies. Are they all shaped like the Milky Way? _____
_____

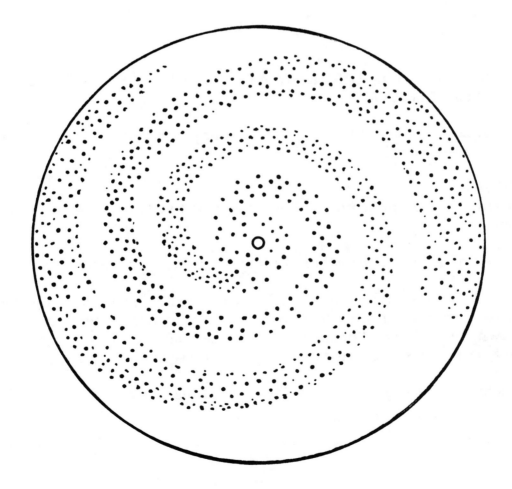

BINOCULARS SUN VIEWER

WARNING: DO NOT LOOK DIRECTLY AT THE SUN, PARTICULARLY THROUGH BINOCULARS OR A TELESCOPE.

WHAT YOU NEED: Binoculars, a piece of cardboard or poster paper, scissors, a compass, typing or drawing paper and a pencil.

WHAT TO DO:
1. Cut one circle in the cardboard to fit over one objective lens on the binoculars. Cover the other lens.

2. With the compass, draw a circle three inches in diameter in the center of the paper.

3. Center the binoculars on the sun. DO NOT look through the binoculars to do this. Hold the sheet of paper so the sun's image is reflected on the paper in the center of the circle you have drawn. Move the sheet of paper so the image completely fills the circle. Trace the dark spots and marks you see reflected on your paper. These are sunspots.

4. Do this activity once a day for several days. Record the date of each drawing. What changes do you see? _____
_____

Name_____

SATELLITES IN ORBIT

WHAT YOU NEED: A magnet, a paper clip, a pair of scissors, a ten-inch piece of string, tape and a pinch clothespin.

WHAT TO DO:
1. Cut out the earth circle at the right and tape it to the magnet.

2. Tie the paper clip to the end of the other piece of string. Cut out the satellite and tape it to the paper clip.

3. Hold the earth magnet still. Slowly circle the magnet with the satellite paper clip, holding the satellite close to the earth. What happens?_____

4. Move the satellite farther from the earth and circle the earth again. What happens?_____

5. Let the satellite spin around the earth under its own momentum. Watch closely. What happens?_____

THINK:
1. What are some of the difficulties scientists might have trying to place a satellite in orbit?
_____

2. What can NASA do with old satellites that are no longer useful and are still in orbit around the earth? _____

3. People have suggested using space as a place to dump our toxic trash. Would this be a good idea? Why or why not? _____

HOW TO FEED AN ASTRONAUT

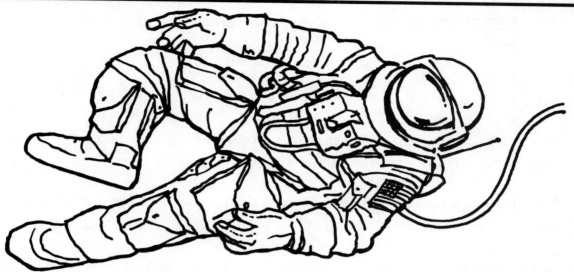

Feeding an astronaut in zero gravity outer space is not an easy task. Liquids do not pour; they don't even stay in a glass. How could you catch coffee that isn't in its cup? Crumbs could be a problem, too. They would float around the cabin until you caught them. Would you want crumbs hitting you in the face as you tried to move around?

For this activity, you need to design a way to feed an astronaut something solid and something liquid without spilling a drop or creating any crumbs. You cannot easily duplicate zero gravity; instead design containers and pick foods that can be eaten and drunk by a person who is upside down.

WHAT YOU NEED:
1. Foods that do not crumb easily or drip. Remember, food such as peanuts do not have crumbs or drip, but they are small. If an astronaut dropped his/her peanuts in outer space, he/she might have a difficult time retrieving them.

2. Containers that prevent dripping and hold the food so it cannot get out.

WHAT TO DO:
1. Design a drinking container that will not leak or drip. Fill it with something to drink.

2. Design a food container that does not allow the solid food to fall out, drip or leak.

3. Have the astronaut hang from a jungle gym or prop the astronaut against a chair, desk or wall so that he or she is upside down. Have him or her test your designs. Do they work?

Name_____

PREPARING FOR THE UNKNOWN

The biggest problem NASA and the astronauts face is how to prepare for the unknown. We don't know everything about our own planet and the life forms that live on it. We know much less about the other planets in our solar system and even less about what is beyond our solar system. The possibilities are endless and the only basis we have for facing the universe is what we do know. What if the laws which work on our planet and the theories which we have developed do not apply in outer space or on planets we have not yet discovered?

If you were an astronaut, how would you prepare for the following:

1. the unknown climate of another planet and the vast temperature changes you might face? _____
_____

2. an atmosphere that contains unknown gases? _____
_____
_____

3. other life forms? _____
_____
_____

4. Use your imagination. Draw and describe a world as different from ours as possible. What would an astronaut need to do to prepare to visit your planet? Use another piece of paper to write about your planet.

Name_____

AN EXPERIMENT TO DO IN SPACE

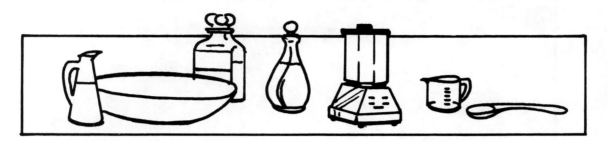

All of our activities on Earth are governed by the law of gravity. If you drop a hammer, a part of your body or your neighbor's body may be in danger. If you salt an egg, the salt drops down and sticks to the egg. If you stir sugar into coffee, it dissolves. All of those things might not happen the same way in space. Some industrial procedures might be easier to perform in outer space where crystals form more easily and uniformly, where weight is not a factor, where there is no "down." Students, scientists and people from industry have devised various experiments to try in space to see how to utilize the effects of zero gravity. The activity that follows is one very small Earth problem: how to keep salad dressing uniformly mixed.

WHAT YOU NEED: Salad oil, vinegar, salad seasoning, a spoon, a measuring cup, a large bowl and a whisk, beater or blender.

WHAT TO DO:

1. Pour one cup of vinegar into the bowl. Add one cup of oil and one teaspoon of spices. What happens? _____

2. Stir the mixture with a spoon. What happens when you stop? _____
_____

3. Beat or blend the mixture until it is thoroughly combined. Let it stand for five minutes. What happens to the mixture? _____
_____

4. Oil is lighter, or less dense, than vinegar; therefore, it floats on top of the vinegar. In outer space, where weight is not a factor, what might happen when you beat or stir this mixture?
_____

CHAPTER SIX

CHEMISTRY

Teacher Notes 116
The Basics: Molecules and States of Matter 116
    Solids, Liquids and Gases 124
    It Doesn't Add Up 125
Chemicals 117
    Precipitates 126
    Mixtures and Solutions 127
Chemical Change 118
    Qualitative Analysis 128
    Acids and Bases 129
    An Endothermic Chemical Change 130
    An Exothermic Chemical Change 131
Nature and Use of Chemicals 120
    Disappearing Dye 132
    Cooking Chemicals 133
    Silver Polish 134
Crystals 121
    Crystal Growing 135
Drugs and the Human Body 121
    Alcohol, Nicotine and Stimulants 136
    Narcotics 137
    Hallucinogens and Cannabis 138
    Inhalants, Depressants and Steroids 139

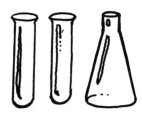 

© Frank Schaffer Publications, Inc.   115   FS-10112 Hands-On Science Experiments

TEACHER NOTES

This section contains some experiments which should not be performed by students without adult supervision. Even household chemicals can be dangerous if swallowed or splashed in the eyes. Where noted, it is advisable to perform these experiments as demonstrations with students assisting and observing. In all cases, do not make substitutions for the chemicals listed in the experiment and heed any precautionary statements which are listed on household chemical labels. For example, ammonia should be kept away from eyes, not ingested, not deliberately inhaled and never combined with chlorine bleaches.

THE BASICS: MOLECULES AND STATES OF MATTER

Models of molecules can easily be made from Styrofoam balls, beads and pipe cleaners. The molecules can be linked with pieces of thin wire or pipe cleaners to demonstrate monomers and polymers or compounds. Use a specific color for each different element and choose bead and Styrofoam ball sizes ranging from eight millimeter beads to four-inch diameter Styrofoam balls to correspond to the atomic weight of the elements. The Styrofoam balls can be painted with tempera or acrylic paints.

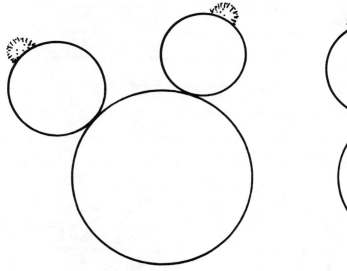
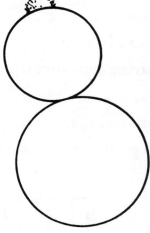

Activity Page 124 Teacher Key

SOLIDS, LIQUIDS AND GASES

This activity explores the states of matter using water as an example. The second half of this activity should be performed as a demonstration by the teacher with the students observing and recording their observations. This is NOT a safe experiment for students to perform individually because the second part requires that the water be heated to the boiling point.

Part 1
   3. The ice will occupy more space than the liquid water and will expand beyond the water level mark.
   4. Ice occupies more space than water.
There are six molecules in a single ring of ice and six sides on a snowflake.

Part 2
   2. When the water boils, it pushes up the aluminum foil lid.
   3. Gas (steam) occupies more space than liquid (water).

Activity Page 125 Teacher Key

IT DOESN'T ADD UP

Part 1
   2. The alcohol and water combined equal less than two cups of liquid.

THINK: The alcohol mixes with the water, and the alcohol molecules occupy the spaces between the water molecules.

Part 2
   1. The oil and water combined equal two full cups of liquid.

THINK: The oil doesn't mix with the water; it floats on top.

CHEMICALS

The chemical elements are the basic building blocks of chemistry. Games provide a way of aiding students in recognizing the elements and associating the correct chemical symbols with those elements. Chemical Dominoes or Chemical Concentration cards can be made from index cards. For dominoes, draw a line down the middle of the cards. Write the chemical name on one end of a card and its correct symbol on the end of another card. Play by the rules for dominoes. For Chemical Concentration, write the name of each chemical element on one card and the symbol on another card. Mix the cards and arrange them face down. When the cards are turned over, one chemical name and its matching chemical symbol constitute a matched pair.

© Frank Schaffer Publications, Inc.

Activity Page 126 Teacher Key

PRECIPITATES

3. A gummy, white precipitate forms. (This precipitate can be probed with a popsicle stick or spoon to determine the consistency.)

THINK: This process could be used to remove pollutants from water, to clean oil, to desalinate water . . .

Activity Page 127 Teacher Key

MIXTURES AND SOLUTIONS

The activity sheet, MIXTURES AND SOLUTIONS, introduces the concepts of mixtures and solutions. Discuss the fact that a solution is a mixture, but a mixture is not necessarily a solution.

1. The salt and the rice can be separated (by hand, using a sieve; the salt could be dissolved or washed out with water; the salt and rice separate when you shake the container).
2. The sugar and water can be separated by boiling the water.

THINK:
1. Mixtures and solutions are both combinations of elements. Neither involves a chemical change.
2. A solution is usually liquid. A substance is dissolved into a liquid. Mixtures are not usually liquid.

_S_ coffee	_M_ jello	_M_ cement	_S_ Kool Aid
_S_ tea	_M_ ice cream	_S_ chocolate milk	
_S_ lemonade	_M_ cake frosting	_M_ mayonnaise	

CHEMICAL CHANGE

Students should be carefully supervised and cautioned before performing these experiments. Some of the chemicals can be dangerous, especially if they are splashed in the eyes. If adequate supervision cannot be maintained, perform the experiments as demonstrations. Under no circumstances should students ingest any of the chemicals used in this section, nor should they eat any of the food which has been tested with iodine.

Activity Page 128 Teacher Key

QUALITATIVE ANALYSIS

1.-6. The spot is a dark purple blue.
7. & 8. The spot is brown.

Paper, laundry starch, potato, bread, apple and flour all contain starch.

Activity Page 129 Teacher Key

ACIDS AND BASES

To test for acids and bases, there are several other test indicators, including an indicator solution to use in place of litmus paper. Litmus paper is easy to use and not messy. This experiment can be performed as a demonstration, but it is also perfectly safe for students to perform individually.

1. The red litmus paper stays red. The blue litmus paper stays blue.
2. The red litmus paper stays red. The blue litmus paper turns red.
3. The red litmus paper turns blue. The blue litmus paper stays blue.
4. The blue litmus paper turns red in the vinegar, then turns blue again when it is dipped in the soap.
5. The red litmus paper stays red. The blue litmus paper turns red.
6. The red litmus paper stays red. The blue litmus paper turns red.

THINK:
1. Vinegar, orange juice and Pepsi or Coke are acids.
2. Milk and liquid soap are bases.

Activity Page 130 Teacher Key

AN ENDOTHERMIC CHEMICAL CHANGE

Use an indoor/outdoor thermometer for the endothermic and exothermic experiments. These experiments will yield more accurate temperature results if they are performed in a Styrofoam cup instead of a glass (the glass conducts heat), but the experiments are more effective as a learning tool when students can see exactly what is taking place. Perform the experiments the first time as they are written; then repeat them using Styrofoam cups and compare the results.

1. The answer should reflect the student's observations.
2. The temperature drops. The final temperature should reflect the student's observations.

THINK:
1. An endothermic chemical change is a chemical change that absorbs heat (or words to that effect).
2. Coolers for drinks, ice packs, ...

Activity Page 131 Teacher Key

AN EXOTHERMIC CHEMICAL CHANGE

1. The answer should reflect the student's observations.
2. The temperature rises. The final temperature should reflect the student's observations.

THINK:
1. An exothermic chemical change is a chemical change that yields heat.
2. Foot and hand warmers, food warmers, heating pads, ...

Chemicals are used in every facet of our lives. Discuss the function of chemicals in medicine and pharmaceuticals, food and drink, the making of plastics, textiles, paper, paints and dyes, perfumes, soaps, fuel, pottery and china glazes. The following experiments are examples of the use of chemicals as well as a demonstration of the way chemicals perform.

NATURE AND USE OF CHEMICALS

Activity Page 132 Teacher Key

DISAPPEARING DYE

Following the experiment, DISAPPEARING DYE, discuss the effect bleach has on clothes. Basically, the stain is camouflaged, not removed from the clothing.

3. The water turns the color of the dye.
4. At first the bleach colors slightly.
5. Gradually the color disappears and the color on the cotton swab also disappears.
6. The water is colored, and the bleach is clear.

THINK: The dye is not gone. It has undergone a chemical change rendering it colorless.

Activity Page 133 Teacher Key

COOKING CHEMICALS

COOKING CHEMICALS is a demonstration of one use of chemicals in the kitchen. Direct the students' attention to the texture of the final product, not the taste. Try varying the amount of baking powder in a muffin recipe for a slightly different texture effect. Discuss the need for chemicals in food preparation and preservation (flavoring, texture, pickling). Point out that with chemicals, as with many other things, there is a law of diminishing returns. More is not always better.

7. The original "Recipe" should be more doughy and more solid. The "x 2" should be slightly lighter and crispier and the "x 4" should be the lightest and crispiest.

Activity Page 134 Teacher Key

SILVER POLISH

4. The tarnish disappears.
5. The aluminum foil is pitted.

THINK:
1. The chemical reaction has removed the tarnish (silver sulfide).
2. Discuss the fact that abrasives rub off and scratch the silver, while the baking soda mixture simply removes the tarnish.

CRYSTALS

Crystal formation, crystal shapes and the natural occurrence of crystals such as rock salt (halite) and the various gem and mineral crystals make a fascinating study. Discuss man-made mineral and gem crystals and their use. Observe different crystal shapes and the size of various crystals.

Activity Page 135 Teacher Key

CRYSTAL GROWING

5. Each crystal has a distinctive size and shape.

THINK: It is cheaper to "grow" some crystals such as diamonds and emeralds than to mine them. Laboratory grown crystals have no defects.

DRUGS AND THE HUMAN BODY

This section is devoted to illegal drugs, addictive drugs and harmful drugs. Invite local medical and law enforcement personnel to speak to the students about drugs and discuss all aspects of drug use and abuse. Have students research topics and then present their findings to the whole class or form two-person teams and have students debate their topics. Also, form panels to present different facets of different drugs and their use and abuse. Use the following topics for discussion and research:

1. Choose two drugs and write a short paper on the effects users of these drugs have on others and their effects on the people around them.
2. Research the lives and deaths of John Belushi, Len Bias and Don Rogers.
3. What are the different methods of treating drug users in different countries? What works? What doesn't work?
4. Which drugs are the most dangerous? Why?
5. How can we prevent drug abuse?
6. How can we cure or rehabilitate addicts?

The four activity sheets in this section cover the various drugs, their common appearances, side effects and dangers. Allow access to library resources to answer all of the questions on these sheets.

Activity Page 136 Teacher Key — ALCOHOL, NICOTINE AND STIMULANTS

1. B   2. C   3. D   4. A

RESEARCH: 1-3. The four drugs listed here range from the highly addictive cocaine to alcohol which is variable. Alcohol and smoking have a detectable odor. Alcohol, cocaine and amphetamines affect behavior. The withdrawal symptoms vary with the drug.

Activity Page 137 Teacher Key — NARCOTICS

1. D   2. C   3. B   4. A

TEST YOUR KNOWLEDGE:
1. Heroin is very addictive. Some people take a longer time to become addicted, but almost all users eventually become addicted.
2. The common side effects of heroin are impairment of mental faculties, inability to concentrate, forgetfulness and confusion. Slurred speech and poor coordination, loss of appetite, slowed respiration and heartbeat, lowered body temperature and flushed skin are the physical effects.

RESEARCH:
1 & 2. Death is the greatest danger but psychosis and mental illness are also dangers. Treatments vary from free clinics to incarceration.

Activity Page 138 Teacher Key — HALLUCINOGENS AND CANNABIS

1. C   2. A   3. B

HALLUCINOGENS RESEARCH: 1-3. These drugs can be highly addictive and mind altering. Call a drug hotline or contact authorities to deal with a "bad trip."

CANABIS RESEARCH: 1-3. These drugs can be highly addictive and physically debilitating. These drugs can have a profound effect on families and friends of users, if only through contact with criminal elements of society.

Activity Page 139 Teacher Key

INHALANTS, DEPRESSANTS AND STEROIDS

1. Some commonly used inhalants are glues and adhesives, paints and lacquers, paint thinners and strippers, dry cleaning fluid, antifreeze, typewriter correction fluid, various aerosols, toluene, gasoline, butane, benzene, dyes, polishes, nail polish remover, etc. The list is almost endless.

2. The dangers of inhalants vary depending on the chemical content. Benzene causes anemia and leukemia, dry cleaning fluid damages the liver and kidneys, toluene damages the brain, leaded gasoline causes lead poisoning, butane damages the heart and many inhalants contain poisons which can result in instantaneous death. In addition, since many users cover their faces with plastic bags to concentrate the drug, suffocation is one of the dangers. Also, many of the inhalants produce a feeling of nausea, and a prone user can easily choke on vomit.

3. The dangers of depressants are psychological and physical dependence and severe and sometimes fatal reactions to sudden withdrawal. In combination with alcohol they can depress breathing, resulting in coma and death. The body tends to build up a tolerance to depressants quickly, making increased dosage necessary to maintain the desired effect. The margin between an effective dose and a lethal dose of these substances is narrow, and death from overdose is one of the leading causes of poisoning deaths in the United States.

4. Steroids can damage the liver and kidneys, heart and other organs. In addition, continued use can increase aggression and violent tendencies, cause acne and baldness and render the user sterile.

THINK: 1 & 2. People take these drugs for many reasons: social, psychological or simply out of curiosity. Drug awareness programs and social pressure are the best deterrents.
For a counselor, for referral or for answers to questions about drug use and abuse, advise students to call this crisis hotline:

National Institute on Drug Abuse
Treatment Referral
1-800-622-HELP
This hotline is open 9 a.m. to 3 p.m. weekdays and 12 noon to 3 p.m. on weekends.

For information about drugs, call:

National Federation of Parents for a Drug Free Youth
1-800-554-KIDS

This is not a crisis hotline. They provide pamphlets, books, videos and information on drugs for children and adults.

© Frank Schaffer Publications, Inc.   FS-10112 Hands-On Science Experiments

Name_____

SOLIDS, LIQUIDS AND GASES

Part 1
WHAT YOU NEED: A clear plastic container, a marker and water.

WHAT TO DO:
1. Fill the container one half full of water. Using the marker, record the level of the water on the outside of the container.

2. Place the container in the freezer overnight.

3. Remove the container from the freezer and check to see the level of the ice in the container. What happened? _____

4. Which occupies more space, liquid water or solid ice? _____
_____

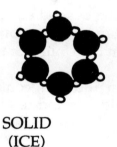  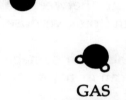

SOLID   LIQUID   GAS
(ICE)   (WATER)   (STEAM)

Look closely at the ice molecules above. How many molecules are there in a single ring? _____ How many sides are there on a snowflake? _____

Part 2
WHAT YOU NEED: A glass beaker, water, aluminum foil and a hot plate or Bunsen burner.

WHAT TO DO:
1. Fill the beaker half full of water. Cover the beaker tightly with the aluminum foil.

2. Watch while your teacher heats the beaker until the water boils. What happens? _____
_____

3. Which occupies more space, liquid (water) or gas (steam)? _____

Name_____

IT DOESN'T ADD UP

One plus one equals two, doesn't it? If you add one cup of liquid to another cup of liquid, doesn't that mean you will have two cups of liquid?

WHAT YOU NEED: Two measuring cups (one must be a clear, glass cup that will hold two cups of liquid), rubbing alcohol, oil, water and food coloring.

Part 1
WHAT TO DO:
1. Measure one cup of water and pour it into the clear, glass measuring cup. Add two drops of food coloring and stir it.

2. Measure one cup of rubbing alcohol and pour it into the clear measuring cup on top of the water. How many cups of liquid do you have in the clear cup? _____

THINK:
Why doesn't one cup of liquid plus one cup of liquid add up to two cups of liquid? (Hint: Look at a diagram of water molecules. Are there any spaces between the molecules?) _____
_____

Part 2
WHAT TO DO:
1. Repeat the experiment you just performed, only this time, use oil in place of the alcohol. What amount of liquid do you have this time?_____
_____

THINK:
Why do you get different results when you use oil instead of alcohol? _____
_____
_____

Name_____

PRECIPITATES

WHAT YOU NEED: Alum, household ammonia, warm water and a pint glass jar.

WHAT TO DO:
1. Pour one cup of warm water in the glass jar.

2. Stir one teaspoon of alum into the warm water until the alum is completely dissolved.

3. Add two teaspoons of ammonia to the solution of alum and water. What happens?_____
_____

What you see in the jar is called a *precipitate*.

THINK:
For what purpose could you use the process of precipitation? _____
_____
_____

Name_____

MIXTURES AND SOLUTIONS

WHAT YOU NEED: One half cup of rice, one half cup of salt, one fourth cup of sugar, one cup of warm water, a spoon and two containers for mixing.

WHAT TO DO:
1. Place the rice and the salt in one container. Stir until they are thoroughly mixed. This is a mixture. Can you separate the salt from the rice now that they are mixed? _____
_____

2. Pour the water into the other container. Add the sugar and stir until the sugar is dissolved. This is both a mixture and a solution. Can you separate the sugar from the water now that they are mixed? _____
_____

THINK:
1. How are mixtures and solutions the same? _____
_____

2. How are mixtures and solutions different? _____
_____

Which of the substances listed below are solutions and which are just mixtures? Place an M in the blank beside the mixtures and an S in the blank beside the solutions.

_____ coffee	_____ jello	_____ cement	_____ Kool Aid
_____ tea	_____ ice cream	_____ chocolate milk	
_____ lemonade	_____ cake frosting	_____ mayonnaise	

Name_____

QUALITATIVE ANALYSIS

Chemists often need to know what substances are in a compound, mixture or solution. The process of analyzing these substances is called *qualitative analysis*. In this experiment, you will perform one test to determine the identity of substances. The substance you will test for is starch.

WHAT YOU NEED: Iodine, a piece of white paper, laundry starch, a slice of potato, bread, a slice of apple, flour, cheese and sugar.

WHAT TO DO:

1. Place a drop of iodine on the white paper. What color is the spot?_____

2. Place a drop of iodine on the laundry starch. What color is the spot? _____

3. Place a drop of iodine on the potato. What color is the spot? _____

4. Place a drop of iodine on the bread. What color is the spot? _____

5. Place a drop of iodine on the apple. What color is the spot? _____

6. Place a drop of iodine on the flour. What color is the spot? _____

7. Place a drop of iodine on the cheese. What color is the spot? _____

8. Place a drop of iodine on the sugar. What color is the spot? _____

Iodine turns a dark purple blue when it touches starch. Which of the items you tested contain starch? _____

Name_____

ACIDS AND BASES

WHAT YOU NEED: Red and blue litmus paper and shallow containers with small amounts of water, milk, vinegar, liquid soap, orange juice and Pepsi or Coca Cola.

WHAT TO DO?

1. Dip a piece of the red litmus paper in the water. What happens? _____

   Dip a piece of the blue litmus paper in the water. What happens? _____

2. Dip a piece of the red litmus paper in the vinegar. What happens?_____

   Dip a piece of the blue litmus paper in the vinegar. What happens?_____

3. Dip a piece of the red litmus paper in the soap. What happens? _____

   Dip a piece of the blue litmus paper in the soap. What happens? _____

4. Dip a piece of the blue litmus paper in the vinegar and then dip it into the soap. What happens? _____

5. Dip a piece of the red litmus paper in the orange juice. What happens?_____

   Dip a piece of the blue litmus paper in the orange juice. What happens?_____

6. Dip a piece of the red litmus paper in the Pepsi or Coke. What happens? _____

   Dip a piece of the blue litmus paper in the Pepsi or Coke. What happens? _____

Litmus paper turns red in the presence of an acid and blue in the presence of a base (alkaline).

THINK:

1. Which of the substances you tested are acids? _____
   _____

2. Which of the substances you tested are bases? _____
   _____

Name_____

AN ENDOTHERMIC CHEMICAL CHANGE

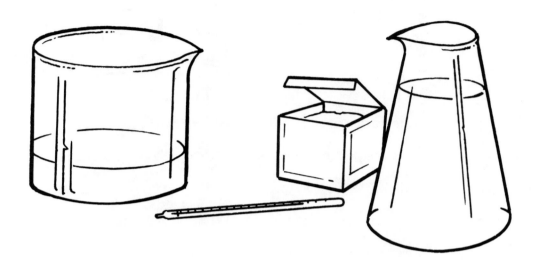

WHAT YOU NEED: Ammonium nitrate, water, a glass beaker and a thermometer.

WHAT TO DO:
1. Pour 100 milliliters of water into the beaker. Insert the thermometer into the water. What is the temperature of the water? _____

2. Dump 15 grams (one level tablespoon) of ammonium nitrate into the water all at once. Watch the thermometer. What happens? _____

   What is the final temperature of the water? _____

THINK:
1. You have just observed an endothermic chemical change. In your own words, what is this?
   _____
   _____

2. For what do you think a chemical change like this could be used? _____
   _____

© Frank Schaffer Publications, Inc.   130   FS-10112 Hands-On Science Experiments

Name_____

AN EXOTHERMIC CHEMICAL CHANGE

WHAT YOU NEED: Powdered bleach, a beaker, a thermometer and water.

WHAT TO DO:
1. Fill the beaker half full of water. Insert the thermometer into the water. What is the temperature of the water? _____

2. Add one tablespoon of powdered bleach. Watch the thermometer. What happens? _____
   _____

   What is the final temperature of the water? _____

THINK:
1. This is an example of an *exothermic* chemical change. In your own words, what is an exothermic chemical change? _____
   _____

2. What could an exothermic chemical change be used for? _____
   _____

Name_____

DISAPPEARING DYE

WHAT YOU NEED: Two clear glass containers, two cotton swabs, water, bleach and food coloring.

WHAT TO DO:

1. Fill one glass container half full of water.

2. Fill the second container half full of bleach.

3. Dip one cotton swab in food coloring and then stir it in the container of water. What happens? _____

4. Dip the second swab in food coloring and then stir it in the container of bleach. What happens? _____

5. Continue to watch the container of bleach. What happens? _____
_____

6. After five minutes, compare the two containers. What has happened?_____
_____

THINK:

Is the dye actually gone? Has the dye disappeared from the container or has a chemical change made it colorless?_____

Name_____

COOKING CHEMICALS

WHAT YOU NEED: Flour, confectioner's sugar, salt, Crisco, baking powder, measuring spoons, three mixing bowls, a measuring cup, three nine-inch pie pans and a stirring spoon.

THE RECIPE:

    1 cup of flour
    1/8 teaspoon of baking powder
    1/8 teaspoon of salt
    1/2 cup of Crisco
    1/4 cup of confectioner's sugar

Bake in an ungreased pan at 350 degrees for 20 to 25 minutes until the edges are lightly browned.

WHAT TO DO:
1. In each bowl, mix 1/2 cup of Crisco and 1/4 cup of confectioner's sugar until the mixtures are creamy.

2. To bowl number one, add 1 cup of flour, 1/8 teaspoon of baking powder and 1/8 teaspoon of salt. Label this bowl "Recipe."

3. To bowl number two, add 1 cup of flour, 1/4 teaspoon of baking powder and 1/8 teaspoon of salt. Label this bowl "x 2" to show that you doubled the baking powder.

4. To bowl number three, add 1 cup of flour, 1/2 teaspoon of baking powder and 1/8 teaspoon of salt. Label this bowl "x 4" to show that you have added four times the amount of baking powder.

5. Preheat the oven to 350 degrees.

6. Pat the dough from bowl #1 into a pie pan and mark the top of the dough with an *R* for recipe. Pat the dough from bowl Number 2 into the second pie pan and mark it with "x 2." Pat the dough from bowl Number 3 into the third pan and mark it with "x 4." Bake all three pans in the oven for 20 to 25 minutes or until the edges turn light brown.

7. Remove from the oven and allow the cookies to cool before eating. What is the difference between the three cookies? _____

© Frank Schaffer Publications, Inc.   FS-10112 Hands-On SCIENCE EXPERIMENTS

Name_____

SILVER POLISH

WHAT YOU NEED: A glass or enamel pan, aluminum foil, baking soda, salt, hot water and a piece of tarnished silver.

WHAT TO DO:
1. Place a sheet of aluminum foil on the bottom of the pan.

2. Place the tarnished silver on top of the foil.

3. Mix one teaspoon of salt and one teaspoon of baking soda into one quart of hot water.

4. Pour the solution over the tarnished silver. What happens? _____
   _____

5. When the silver is clean, remove it from the water. Look at the aluminum foil. What has happened to it? _____
   _____

THINK:
1. What happened to the tarnish on the silver? _____

2. Some silver polishes are abrasive; they rub or scratch the tarnish off the silver. Would the baking soda mixture be a better polish for the silver? Why or why not? _____
   _____

CRYSTAL GROWING

Name_____

WHAT YOU NEED: Sugar, alum, salt, epsom salts, hot water, four clear, shallow containers and one beaker or glass jar for mixing.

WHAT TO DO:
1. Pour 1/2 cup of hot water and as much sugar as you can dissolve in the water into the beaker or jar. Pour the liquid into one of the shallow containers and label it "sugar."

2. Rinse the beaker and refill it with another 1/2 cup of hot water. This time, dissolve as much alum as possible in the water. Pour the liquid into a shallow container and label it "alum."

3. Rinse the beaker and refill it with a third 1/2 cup of water. Dissolve as much salt as possible in this water. Pour the liquid into a shallow container and label it "salt."

4. Rinse the beaker and refill it with a fourth 1/2 cup of water. Dissolve as much epsom salts as possible in this water. Pour the liquid into a shallow container and label it "epsom salts."

5. Place all four containers in a dry sunny place. Observe the containers daily. When the water has evaporated, draw the different crystal shapes below. Label each crystal shape with the correct name. Can you tell the difference between substances by their crystal shape? _____
_____

6. In the space at the bottom of the page, draw and label each of the four crystals you have grown.

THINK:
Why is it useful to be able to "grow" certain crystals?

Name_____

ALCOHOL, NICOTINE AND STIMULANTS

Different drugs have different effects on the human body. Each drug has both short term effects and long term effects, and many of these effects can be dangerous or fatal. Match the drug below with its potential dangers. Write the letter of the correct list of dangers next to the drug with which it belongs.

DRUGS

_____ 1. Nicotine (tobacco)

_____ 2. Alcohol

_____ 3. Amphetamines (stimulant)

_____ 4. Cocaine (stimulant)

A. Physical dangers are damage to nose tissues, insomnia, weight loss and heart irregularities including increased chance of cardiac arrest. Mental dangers are cocaine psychosis (this condition is similar to paranoid schizophrenia; the addict thinks bugs are crawling under his skin), paranoia, severe depression and hallucinations.

B. Physical dangers are lung cancer; cancer of the larynx, bladder, tongue, cheek and gums; emphysema; cardiovascular diseases; strokes and atherosclerosis (the buildup of fat along the walls of blood vessels).

C. Physical dangers are cirrhosis of the liver (the normal liver tissue is replace by an inflexible tissue that impairs function; if the drinker continues to drink heavily, this condition results in death). Mental dangers are depression, aggression, impaired coordination, and loss of self-control.

D. Physical dangers are dehydration, fever, sleeplessness and high blood pressure (and the risk of heart failure). Mental dangers are amphetamine psychosis (hallucinations and delusions), paranoia, and violent and bizarre behavior.

RESEARCH: Use the books in the library to find the answers to these questions. Write your findings on a separate piece of paper.

1. How would you know if someone were using these drugs?

2. How addictive are these drugs?

3. What are the withdrawal symptoms of these drugs?

Name_____

NARCOTICS

What are these drugs? Where can you find these drugs? Match the drug below with its correct description. Write the letter of the correct description next to the drug with which it belongs.

_____ 1. Opium

_____ 2. Morphine

_____ 3. Codeine

_____ 4. Heroin

A. This is the strongest narcotic. It is usually a white, off-white or brown powder and has no odor. It can be dissolved in water. It can be sniffed or injected and is highly addictive.

B. This is a white powder or tablet that can be mixed with many liquids. It can be drunk or injected or swallowed as a tablet. It is much weaker than the other narcotics and is therefore less frequently used.

C. This is the main ingredient in opium and is usually a light brown or white powder. It is a very effective painkiller and can be sniffed, taken orally or injected.

D. This is a natural narcotic made from the opium poppy. It is usually in a powder form or a sticky, bitter-tasting bar which can be either swallowed or smoked.

TEST YOUR KNOWLEDGE:
1. How addictive is heroin? _____
_____

2. What are the common side effects of heroin? _____
_____
_____

RESEARCH: Go to the library to find the answers to these questions. Write your answers on another piece of paper.

1. What are the dangers of the four drugs listed above?

2. What treatments are available to drug users who wish to be free of their addiction to these drugs? How are the treatments in the United States different from those in other countries?

Name_____

HALLUCINOGENS AND CANNABIS

HALLUCINOGENS: These drugs alter the emotions, thought processes and perceptions of the drug user. Known as the psychedelic drugs, they can produce illusions and hallucinations. Match the drug listed below with its correct description. Write the letter of the correct description next to the drug with which it belongs.

_____ 1. LSD

_____ 2. PCP

_____ 3. Mescaline

A. Technically known as phencyclidine, this drug is known on the street as "angel dust," "hay" or "killer weed." It is usually a white powder that is easy to dissolve in water. It can be smoked, inhaled or injected.

B. This drug comes from the peyote plant. It is usually a white powder that is bitter to the taste and is usually eaten. The tops of the peyote are also sometimes dried and eaten or boiled in water so the liquid can be drunk.

C. This drug is known scientifically as D–lysergic acid dietheylamide. This drug is a colorless, tasteless and odorless liquid. It is swallowed.

RESEARCH: Look in the library for the answers to these questions. Write the answers on a separate piece of paper.

1. What are the effects and dangers of these drugs?

2. What should you do if someone you know has taken one of these drugs and is having a "bad trip?"

3. How addictive are these drugs?

CANNABIS: Marijuana and hashish both come from the cannabis plant. These drugs tend to be psychologically addictive rather than physically addictive. Use your library resources to answer these questions about hashish and marijuana.

1. What are the effects of marijuana and hashish? How would you recognize someone who was using those drugs?

2. What are the physical and psychological dangers of these drugs?

3. What effect do these drugs have on the children of the users and on the families and friends of the users?

Name_____

INHALANTS, DEPRESSANTS AND STEROIDS

Use your library resources to answer these questions.

1. An inhalant is any substance that gives off a vapor which can be inhaled or breathed into the lungs. What are some commonly used drug inhalants? _____
   _____
   _____

2. What are the dangers of using inhalants? _____
   _____
   _____

3. A depressant is usually used to sedate a person or help them sleep. Barbiturates and tranquilizers are the two common forms of depressants. What are the dangers of these drugs? _____
   _____

4. Steroids are often taken to increase strength and athletic ability. What are the dangers of taking steroids? _____
   _____

THINK: Write the answers to the next two questions on a separate piece of paper.
1. Why do people take these drugs?
2. How can we persuade people not to take these drugs?

CHAPTER SEVEN

PHYSICS

Teacher Notes 141
Force 141
    Force 146
    Hot Air Mobile 147
    Air Pressure and Suction 148
Fluid and Density 142
    Oil and Water 149
    Paint Patterns 150
Magnetism 143
    Poles 151
    Static Electricity 152
    Electromagnet 153
Light and Optics 145
    Light Bending 154
    Colors 155

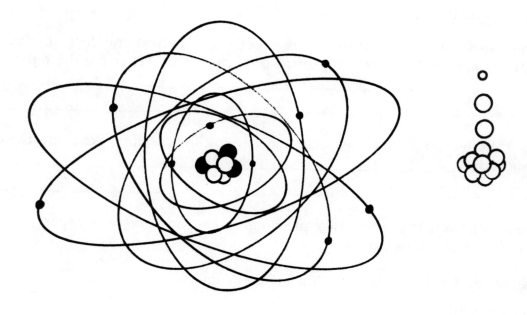

TEACHER NOTES

FORCE

The first activity in this section is a simple demonstration of force. A more solid stand can be built for the apparatus if desired. Paper tubes are stable, economical and easy to prepare in the classroom. The key to making this apparatus work is to string the beads in a straight line an equal length from the dowel. The threads should hang perpendicular to the floor, and the beads should just touch. Discuss the laws of force prior to doing the experiment and ask students for predictions of the results prior to performing steps five and six.

Activity Page 146 Teacher Key — **FORCE**

5. When one bead is pulled away and released, one bead is pushed away at the opposite end.
6. When two beads are pulled away and released, two beads are pushed away at the opposite end.

The second and third activities demonstrate the force of moving air and the principles of air pressure and suction. The mobiles are fragile, so caution students to be careful and to take their time cutting them out and stringing them. Be sure to have extra foil on hand for second attempts.

Activity Page 147 Teacher Key — **HOT AIR MOBILE**

6. The mobile revolves slowly as the warm air rises. (It will revolve faster over hotter heat sources.)

These two additional mobile patterns will also work when enlarged for student use.

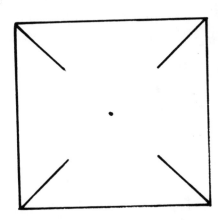

Activity Page 148　Teacher Key

AIR PRESSURE AND SUCTION

Part 1
5. No. Air pressure will hold the water in the straw.

Part 2
3. The cups should attach to the sides of the balloon. They are held in place by suction.

FLUID AND DENSITY

Activity Page 149　Teacher Key

OIL AND WATER

The experiment OIL AND WATER demonstrates the layering effect that is produced when liquids of different densities are poured into the same container. Coloring the water in the third glass helps to make the water and alcohol separation more visible. It does not affect the outcome of the experiment.

1. The oil floats on top of the water.
2. The oil rises above the water and floats on top of it.
4. The alcohol floats on top of the water.
5. The oil forms bubbles and still floats on top of the water.
6. The alcohol and the water mix. Their densities are very similar and the alcohol therefore mixes easily with the water.

THINK:
Accept any reasonable answer. For example, oil spills would be much more difficult to clean up if the oil mixed with the water, salad dressings would be easier to mix if the densities were similar, oil and grease fires would not spread when sprayed with water.

Activity Page 150 Teacher Key

PAINT PATTERNS

PAINT PATTERNS is a rather messy activity. Advise participants to wear smocks or old clothing. Use only two or three colors of paint at a time; they tend to mix after a few minutes in the water because they are water soluble. The paint needs to be quite thin and fluid, and some paint colors work better than others. The metallic acrylic paints tend to sink to the bottom and are difficult to use for this activity. Food coloring can be substituted for the paint and requires no diluting. Add only a drop or two to the water. The food coloring also stains.

THINK:
The textile industry might use this process for dying cloth; it might be used for dying notepaper, book binding paper, wrapping paper, wallpaper; it could be used in plastics and rubber products to make such things as balls and balloons.

MAGNETISM

Activity Page 151 Teacher Key

POLES

The activity, POLES, is a simple demonstration of the patterns of magnetic force fields. Use this as an introduction to magnetism and discuss the value of each kind of magnet. For instance, a horseshoe magnet would not be usable as a compass. Test the force of a bar magnet versus a horseshoe magnet with a flexible magnet strip.

3. The force field and the patterns of the filings are different for each kind of magnet.

Appropriate patterns are provided below.

BAR MAGNET   ROUND MAGNET   HORSESHOE MAGNET

ELECTRICITY

Activity Page 152 Teacher Key

STATIC ELECTRICITY

Use the experiment STATIC ELECTRICITY to introduce positive and negative charges. Relate this to the concept of magnetic poles. The experiment ELECTROMAGNET also uses the concept of positive and negative charges and demonstrates the poles of a magnet as well.

3. The two balloons hang parallel to each other without moving.
6. The balloons move together. The positive and negative charges attract each other, and the balloons are pulled together.
8. The balloons hang side by side, touching each other.
9. The plastic wrap will give both balloons a negative charge.
10. The like charges repel each other, and the balloons fly apart.

Activity Page 153 Teacher Key

ELECTROMAGNET

Review the concepts of charges and poles before performing this experiment. Make certain that students understand that a compass is also a magnet and that the earth's poles are magnetic poles. Following this experiment, discuss the ways electromagnets can be used and the value of electromagnets.

4. The compass needle turns to the magnet.
5. The compass needle turns the away from the magnet.

THINK:
1. Accept any reasonable answer. (Discuss the industrial uses for electromagnets: lifting heavy objects, stirring chemicals and paints, used as parts in motors, etc.)
2. Electromagnets can be turned off.

LIGHT AND OPTICS

Demonstrate the use of a magnifying glass and show students various types and sizes of lenses: convex, concave, meniscus (convex on one side and concave on the other), plano-convex (flat on one side and convex on the other) and plano-concave (flat on one side and concave on the other). Check with a local optometrist's office for old eyeglass lenses to use as examples.

Discuss the way that a glass (or water) refracts or bends light as the light passes through it. Demonstrate the bending of light with a clear glass full of water. Objects behind or inside of the glass have a different appearance.

To make a simple lens, cover a mason jar ring loosely with clear plastic wrap. Secure the wrap with a rubber band, allowing the wrap to dip slightly in the center of the ring. Fill the dip with water. The result is a plano-convex or magnifying lens.

Before performing the experiment LIGHT BENDING discuss the difference between the way a prism bends light and the way a concave or convex lens bends light.

Activity Page 154 Teacher Key

LIGHT BENDING

1. Violet, blue, green, yellow, orange and red are the colors in a rainbow.

THINK:
3. The image produced by sending light through a convex lens is upside down. (Relate this to the lens in the human eye and note that our brain translates the picture to right side up.)

Activity Page 155 Teacher Key

COLORS

The experiment COLORS demonstrates the colors that are in light in reverse. This is a two-page activity with the patterns on the second page. Reproduce the patterns on heavy drawing paper or mount them onto poster board for durability. Have students color the disks using tempera, markers or crayons (Crayola blue-green is the perfect turquoise color for the rainbow spinner). The colors must be heavy and bright enough to cover the writing on the patterns. Red, blue and green in alternate sections on a wheel will also produce white, but the whole rainbow is given here to relate the experiment to the entire spectrum of sunlight.

3. The red and green make a brownish-orange color.
4. The red and blue make a purple color.
5. The blue and green make a blue-green or turquoise color.
6. The rainbow disk makes white.

© Frank Schaffer Publications, Inc.

Name_____

FORCE

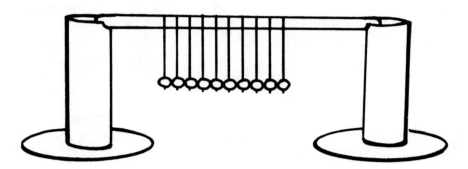

WHAT YOU NEED: 2 identical tubes from paper towel rolls, 2 cardboard disks 5" in diameter, 1/4" diameter dowel, 10 wooden or plastic beads 3/4" to 1" in diameter, scissors, tape and heavy thread.

WHAT TO DO:

1. Cut a small semicircle out of one end of each paper towel tube. Tape the cardboard disks to the opposite end of each roll to make a stand for the roll.

2. Place the dowel in the semicircular cuts of the towel tubes. This forms a swing stand.

3. Cut 10 pieces of thread each one foot long. Make certain the threads are the same length.

4. Tie one end of each thread to a wooden bead. Tie the other end of the thread to the dowel. Keep the beads hanging the same distance from the dowel and place the threads on the dowel so the beads touch one another and the threads hang straight up and down.

5. When the bead swing is assembled, steady the beads to keep them from moving. Pull one end bead out from the other beads and let it go. How many beads move away from the opposite end of the row of beads?_____

6. Pull two beads out from one end and let them go. How many beads move away from the opposite end? _____

© Frank Schaffer Publications, Inc.

Name_____

HOT AIR MOBILE

WHAT YOU NEED: Heavy aluminum foil, scissors, posterboard, scissors, a lamp, a needle and heavy thread.

WHAT TO DO:
1. The patterns for this mobile are at the bottom of this page. With a pencil, lightly trace the first pattern on a piece of heavy aluminum foil.

2. Trace the central disk pattern onto the posterboard.

3. Cut out the pattern pieces and cut on all lines, being careful not to cut beyond the lines.

4. Thread the needle with a 24-inch piece of thread. Double the thread and knot it. Pierce the dot on the central disk with the needle and pull the thread through the hole, being careful not to pull the knot through.

5. Pierce the dot on the mobile and pull the thread through the hole until the mobile rests on the central disk.

6. Cut the needle off the thread. Hold the mobile above a lighted lamp or a heat source. What happens? _____
_____

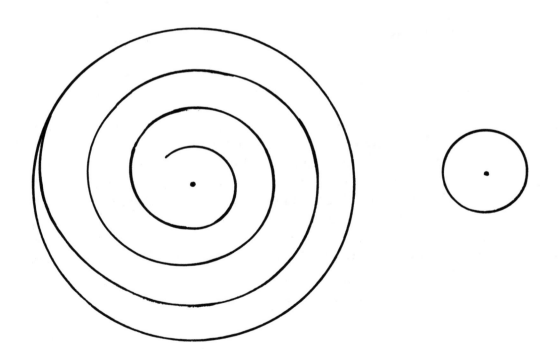

© Frank Schaffer Publications, Inc.   FS-10112 HANDS-ON SCIENCE EXPERIMENTS

AIR PRESSURE AND SUCTION

Part 1: Air Pressure
WHAT YOU NEED: A clear plastic straw, tape, a 6" thread, 1/2 cup of water, food coloring and an eyedropper.

WHAT TO DO:
1. Mix one drop of food coloring in the water.

2. Knot the thread and bend the straw around the thread like a bridle on a hobby horse.

3. Tape the straw flat to hold the string in place.

4. Turn the straw upside down and use an eyedropper to fill the straw with colored water. Fill it to the top.

5. Hold your finger over the open end of the straw and turn the straw right side up. Hold the string by the thread and remove your finger from the end of the straw. Does the colored water drain out of the straw? _____

Part 2: Suction
WHAT YOU NEED: A nine-inch balloon, two six-ounce paper cups and a partner.

WHAT TO DO:
1. Have your partner hold the two paper cups in the air with the open ends pointing toward each other and a four-inch space between the cups.

2. Inflate the balloon in the four-inch space between the cups. As you blow into the balloon, the sides should press against the open ends of the cups and form a seal.

3. Hold the end of the balloon so that no air can escape and ask your partner to let go of the cups. What happens? _____

Name_____

OIL AND WATER

WHAT YOU NEED: Three clear drinking glasses, water, rubbing alcohol, salad oil, two colors of food coloring, two mixing containers and a measuring cup.

WHAT TO DO:
1. Pour 1/2 cup of water into the first glass. Add 1/4 cup of salad oil to the glass. What happens?_____

2. Pour 1/4 cup of salad oil into the second glass. Add 1/2 cup of water to the glass. What happens?_____

3. In one mixing container, combine 1/2 cup of water and one drop of food coloring. In the other mixing container, combine 1/2 cup of rubbing alcohol and one drop of a different color of food coloring.

4. Pour the colored water into the third glass. Tipping the glass sideways, slowly add the colored rubbing alcohol by pouring it down the inside of the glass. Stand the glass upright and observe. What happens? _____
_____

5. Stir glasses one and two. What happens? _____
_____

6. Stir the third glass. What happens? _____
_____

THINK:
1. When would the density of oil and water be important? In what situation would it be helpful that oil is lighter than water?_____
_____

Name_____

PAINT PATTERNS

WHAT YOU NEED: Acrylic paints, a 9" x 13" rectangular baking pan, typing or drawing paper, a spoon or popsicle stick with which to stir, containers for mixing, newspaper and water.

WHAT TO DO:
1. Mix your paint colors in the mixing containers. Squeeze 1/2" from the tube of paint and add one tablespoon of water. Stir until thinned to the consistency of cream. Mix a different color in each container.

2. Pour water into the baking pan to a depth of 1/2".

3. Slowly and carefully, pour the paint mixtures on top of the water. With a clean stick or spoon, very gently swirl and mix the colors to form a pretty pattern.

4. Lay a sheet of clean paper on top of the paint in the pan. Lift the sheet out and lay it on newspaper to dry.

THINK:
1. What uses could you have for this process? What industries might utilize this process? ____

© Frank Schaffer Publications, Inc.

Name_____

POLES

WHAT YOU NEED: Paper, iron filings, a bar magnet, a round magnet and a horseshoe magnet.

WHAT TO DO:
1. Place the magnets on a flat surface. Move them far enough apart so that you can cover each one with a separate piece of paper.

2. Sprinkle iron filings on each piece of paper over the magnets.

3. Draw the magnets and the patterns which the filings make around the pole of each of the magnets. Are they different? _____
_____

BAR MAGNET	ROUND MAGNET	HORSESHOE MAGNET

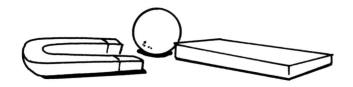

© Frank Schaffer Publications, Inc.   151   FS-10112 Hands-On Science Experiments

Name_____

STATIC ELECTRICITY

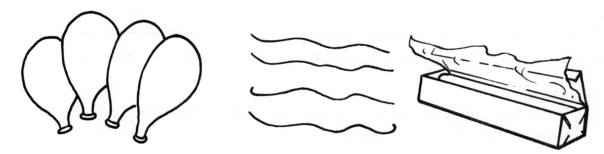

WHAT YOU NEED: Four balloons, nylon thread and clear plastic wrap.

WHAT TO DO:
1. Cut four pieces of thread 18" long.

2. Inflate the two balloons and tie a piece of the thread on each balloon.

3. Hold the two balloons side by side and three inches apart. What happens?_____
_____

4. Rub one balloon on your hair. This will give the balloon a positive charge.

5. Rub the other balloon with the plastic wrap. This will give the balloon a negative charge.

6. Hold the two balloons side by side and three inches apart for a second time. What happens?
_____

7. Inflate the other two balloons and tie a piece of thread on each balloon.

8. Hold the balloons side by side, touching each other. What happens? _____
_____

9. Rub both balloons with plastic wrap. What charge will be on the balloons? _____
_____

10. Hold the two balloons side by side, touching each other for a second time. What happens?
_____
_____

© Frank Schaffer Publications, Inc.

ELECTROMAGNET

WHAT YOU NEED: Four feet of 20 gauge copper wire, electrical tape, a 1.5 volt battery, a compass and a pencil or pen.

WHAT TO DO:
1. Wind the copper wire around the pencil like a spring, leaving six inches free at either end.

2. Using two small pieces of electrical tape, fasten one free end of the copper wire to one terminal of the battery and one end of the wire to the other terminal.

3. Place the compass on a flat surface. Turn the compass until the needle is aligned to point directly north.

4. Point the copper coiled pencil east to west and hold it directly over the compass. What happens? _____
_____

5. Point your pencil magnet in the opposite direction. What happens to the compass needle?
_____

THINK:
1. What practical uses could be made of an electromagnet? _____
_____

2. Why would an electromagnet be more useful at times than a regular magnet? _____
_____

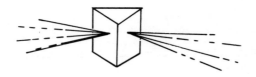

Name_____

LIGHT BENDING

WHAT YOU NEED: A prism, a convex lens, white paper and sunlight.

WHAT TO DO:
Sunlight is composed of different colors of light which separate into a "rainbow" when they pass through a prism.

1. Hold the prism so that the light from the sun shines through it onto the white paper. Tilt the prism different ways if you cannot get a "rainbow" on the paper. What colors are in the rainbow? _____
   _____

2. Draw the rainbow the prism has produced at the bottom of this page.

THINK:
A convex lens bends light a different way. The prism created a rainbow on paper. What do you think a lens would produce? _____
_____
_____

3. Hold the lens near a window so that the light can shine through the lens onto a sheet of white paper. Move the paper back and forth until you get a sharp picture or image on the paper. What sort of image do you have? _____
   _____

© Frank Schaffer Publications, Inc.   FS-10112 Hands-On Science Experiments

COLORS

Name_____

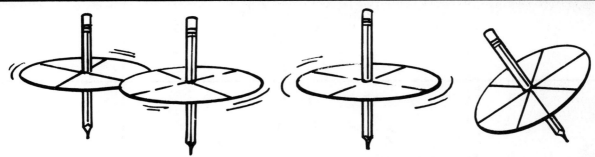

WHAT YOU NEED: Tempera or acrylic paints or colored markers; scissors; a piece of paper and a short, sharp pencil.

WHAT TO DO:

1. Cut out the circles on the second page and color them exactly as they are labeled. If the color name is not written outside the space, leave the space white. Make the colors very solid and bright.

2. Pierce the center of the first disk (red and green). Insert the pencil point in the hole and slide the disk part way up the pencil, with the colors facing the eraser end of the pencil. Do not twist the disk; it needs to be securely attached to the pencil to keep it from sliding off. Tape it to the pencil on the underside if necessary.

3. Place the point of the pencil on the piece of paper and spin the pencil like a top. What happens? _____
_____

4. Remove the disk and replace it with the second disk (red and blue). Repeat steps two and three with this disk. What happens when you spin the second disk? _____
_____

5. Remove the disk and replace it with the third disk (blue and green). Repeat steps two and three with this disk. What happens when you spin the third disk? _____
_____

6. Remove the disk and replace it with the fourth disk (the rainbow disk). Repeat steps two and three with this disk. What happens when you spin the fourth disk? _____
_____

© Frank Schaffer Publications, Inc.   FS-10112 Hands-On Science Experiments

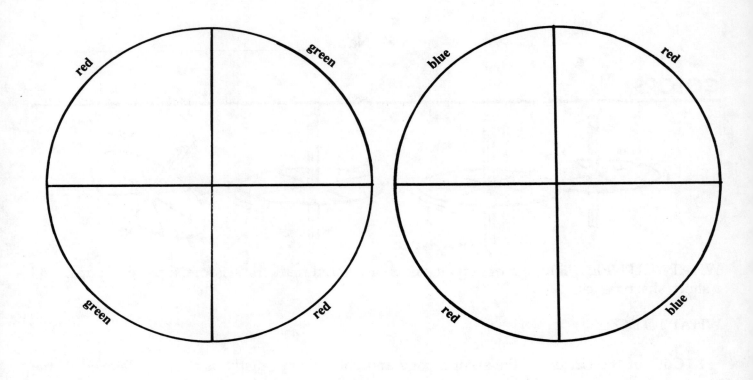
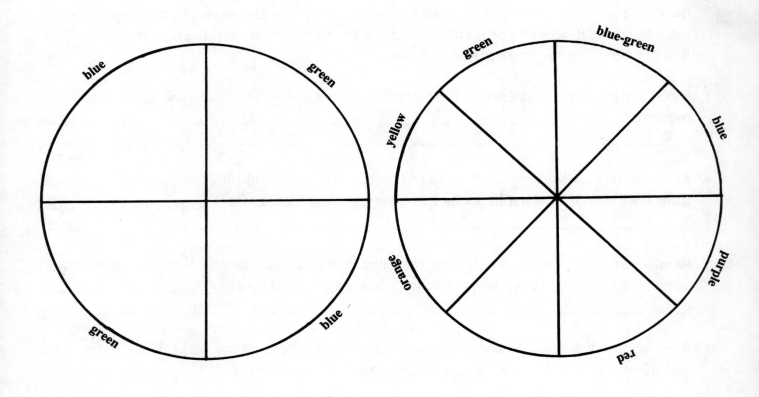